A *Bit* of an

How the Institute of Patent
has helped inventors for 80 years

Derek Myhill

IPI Press

First published in the United Kingdom by IPI Press.

IPI Press
The Institute of Patentees and Inventors,
Suite 505a, Triumph House,
189 Regent Street,
London
W1R 7WF

Telephone: 020 7434 1818
Facsimile: 020 7434 1727
www.invent.org.uk

Cover Design by Spagetti Media,
Telephone 01403 266107

Printed in Great Britain by First Feature Publishing and Printing Limited,
Telephone: 023 8058 4696

British Library Cataloguing in Publication Data.
A Catalogue record of this book is available from the British Library.

ISBN 0-9537420-0-8

Dedicated to all those inventors who have 'a Bit of an Idea'.

Being an inventor, and having successfully built businesses from inventions and new ideas, it gives me great pleasure to be able to give something back to the inventive community by sponsoring this book and the Institute's 80th birthday celebrations.

Britain's wealth is driven by our inventive spirit and it is imperative that this national treasure is encouraged and nurtured by industry and Government.

I salute the efforts of all the Institute's members over the last 80 years and look forward to seeing their wonderful inventions in the future.

Paul Davidson
Davidson Innovation

DEREK MYHILL, MSc, FRSH, FInstPI.

Derek is now retired after a career in the Biomedical Sciences, a Life-Scientist. He is from Manchester and spent early years in research on Rheumatism, at Medical Schools in Manchester, Vermont and Oregon. In 1955-7, he was a lab technologist in the RAMC in Kenya, and speaks Swahili. Whilst at the Oregon Medical School, he was awarded Honorary Citizenship of Portland for scholastic achievement. In 1968 he was Technical Director of the New Orleans Central Pathology Labs. From 1972 he became Director of Chemistry at Oxford Laboratories Inc, in California. He was also a Lecturer in Analytical Toxicology at the UCSF. Back in the UK from 1975, Derek was Consultant to Pharmaceutical companies. During the 80s, traded as 'Bioref', 'Brew-Sciences' and then Sigma Games Ltd. Other inventors gravitated to him with their problems. By 1991, he had founded the IPDC, a self-help group for inventors in the Wessex region. He is co-author of a number of comprehensive reference books on inventors/innovation, now ready for publication.

ACKNOWLEDGMENTS:

Ruth Magnus, the Institute Secretary, provided many relevant documents. She frequently had her day unravelled by my visits and activities. But thanks also go to her for coming up with a name for the book. Ruth says many phone enquiries start with "I've had a Bit of an Idea", then there is a half-hour explanation. **Paul Ambridge**, the Institute's Chairman, helped me get started and gave guidance on pictures/illustrations for scanning and reproduction.

I used the *'INVENTORS WORLD'* magazine's offices as a communications base. I am grateful to **David Wardell** and **Peter Knight** for this facility. Peter has efficiently coordinated all material prior to printing, and arranged the design of the book cover. **Nicholas Mark**, in Newcastle, has been kind enough to do the scanning of all the pictures for the book. This has naturally meant preparing far more pictures than have actually been used, so it has been a lot of work.

Other assistance appreciated has been proofreading by **Graham Jones** and **Douglas W. Nield**, plus Ruth, Paul and especially Peter (final material). **Mike Roos** provided insight into the Ron Hickman saga. **Mr Pat Russell-Rayner, Max Beaumont** and **Perry Akhurst** helped clarify management sequences for the 80s.

However, without the financial support of Mr Paul Davidson it would not have been possible to publish this history in book form. We thank him for the opportunity to present the story in an attractive style suitable for the public too.

CONTENTS

PICTURES/ILLUSTRATIONS PLATES

inserted after page 32

INTO A NEW MILLENNIUM!

A history of the IPI is long overdue given its long and industrious past down the years of the twentieth century, and yet who could have thought that the story of this hard working organisation would turn out to be so fascinating and illuminating. My sincere congratulations go to the author Derek Myhill who's dedicated research has unearthed glittering tales of often unsung inventors working together and swapping ideas whilst trying to further the cause of the Institute. I know that you will thoroughly enjoy every page.

The irony of the story is that very little has really changed in nearly a century in terms of the colossal mountains that inventors are expected to climb in order to get their voices heard and their ideas onto the market place.

It is inexcusable to ignore such enormous potential for Britain's economy and yet successive governments have failed totally to understand how the difficulties of the lone inventor are very different to those of research and development departments of large corporations and those of small businesses. Businesses can stand up for themselves by and large, whereas the lone inventor needs protection and advice from the moment their idea is born, largely because by definition they are not businesses and often have no background in business. Governments including the current one don't truly seem to understand this critical difference. The result is that not nearly enough British inventions are successful through no fault of the inventor and despite the hard work of members of the IPI. Paul Ambridge, the Chairman of the IPI has some truly frightening figures about what this failure means to the British economy, let alone the utter frustration and economic disadvantage to the inventor! We spend a lot of time arguing this case with the DTI in order to persuade them to work with us more closely, and if one thing ensures a long future of the Institute it is the determination to finally win this argument!

But if the core problems faced by inventors remain the same then so too do the benefits of being a part of the IPI. Nowhere else can inventors get impartial help and advice with no other agenda than to steer the inventor towards success along what we all know is an extremely hazardous and rocky road. The philosophy of the Institute is exactly the same today as it was eighty years ago. It is the collective experience of members that is the true value of the organisation, experience that is shared and passed on to other members freely. Although the Institute boasts experts in many fields including patent law and intellectual property rights, the IPI mainly listens to its hundreds of talented individuals and puts together sound advice that is hopefully relevant to everyone. In the past it was perhaps easier to listen. As this book reveals, the IPI once had it's own club and even a bar where members could meet, swap stories and develop ideas. Now that's one piece of our history that I would personally wish to repeat itself before we get too far into the next millennium. As you see, there's still a lot to work towards!

Howard Stableford
President, IPI.

FOREWORD

The Institute of Patentees and Inventors receives thousands of telephone calls each year from people who have "**a *Bit* of an Idea.**" Over the past eighty years the Institute has probably helped around 1/4 million people to either turn their 'Bit of an Idea' into a potentially successful and commercial invention or to realise that their idea is not new but is in fact already available in the shops.

Many of the ideas of members of the Institute have turned into successful products generating wealth for both the inventor and the country. How many of us have used a Black & Decker **WORKMATE ®**, warmed ourselves in front of a **MAGI-COAL** effect fire, climbed a **FOLDAWAY** loft ladder, and looked out on the world through **FLOAT GLASS**. All of these products, and hundreds more, have been the result of the inspiration and hard work of members of the Institute.

Many more thousands of ideas have never seen the commercial light of day as the success rate of turning an idea into a marketed product is less than 0.2 %. That means that for every marketed product, over 600 ideas fall by the wayside never to be taken up by industry.

It is a tragic waste of the UK's inventive capabilities that so many good ideas are not taken up by UK industry. Each year around 4,000 UK patents are granted to lone inventors and only 80 marketed products are produced from these. And of those 80 products, overseas rather than UK manufacturers introduce nearly two-thirds of them

A statistic often quoted is that each year, the UK loses £156 billion pounds in intellectual property rights. This figure has been taken from a report produced by MITI (the Japanese equivalent of our DTI) and includes many factors that are erroneous. If we say, being extremely optimistic, that the figure is 99% incorrect, it still leaves the alarming loss as £1.56 billion pounds each and every year!

If UK industry was more open minded to the ideas of its national inventors and if UK inventors became more commercially aware then I firmly believe that the £1.56 billion loss could be stopped and the UK's manufacturing industry regenerated.

The Institute has always tried to change the media portrayal of lone inventors as "nutty professors working in their garden shed". There are some inventors who could be termed nutty professors and many of them work in their garden sheds.

However, 99% of this country's inventors are down-to-earth people who have solved a problem and believe that their solution could be of benefit to others.

They are people from all walks of life and levels of education, but each has the over- riding belief that their idea could be manufactured, marketed and create wealth for them and the country. The tenacity, drive and self-sacrifice exhibited by these lone inventors does not merit the disregard shown to them by industry. Who knows whether the next person with '*a Bit of an Idea*' will be another Sir Christopher Cockerell - **Hovercraft**, Professor Eric Laithwaite - **Linear Electric Motor**, Frank Whittle - Jet Engine, or other well known inventor with a world beating product.

Through the next millennium, the Institute of Patentees and Inventors will be there to help all its members and the public realise their dreams for their bits of ideas.

Paul Ambridge,

Chairman, Institute of Patentees and Inventors.

If you require further details concerning inventions and inventors featured in this book, or have any suggestions or additional information of interest to the Institute, please contact the IPI marking your letter "A Bit of an Idea".
(Contact details for the Institute of Patentees and Inventors are shown on page 80).

THE HISTORY OF THE INSTITUTE

Mr Morgan Morgan, KC

Major anniversaries have traditionally been celebrated by hosting a grand conference and inventions exhibition followed by a dinner. Detailed reports of these activities were then published together in a single special edition of ***'THE INVENTOR'*** magazine, which constituted a valuable souvenir of events. The Silver and Diamond Jubilee publications were notable in this respect. These particular anniversary years were hectic for the management, council and staff.

However, like all great organisations, the Institute had its ups and downs. Previous efforts to compile the history met with little success due to missing archive materials. It was decided that this must be done for the 80th anniversary (1999), before the centennial arrives. Consequently a researcher has been analysing all records available in-house. Other sources of information examined have been at **The British Library, The Patent Office Library, Westminster Archives Museum**, and in records of older members and families in which successive generations have been members. Much of the early historical data has been incorporated into electronic files, making the search easier. The only surviving copies of the Institute journal ***'THE INVENTOR'*** for 1932-1939 are those at the **British Library**. No journals were published between July 1939, and February 1948.

Much of the **'lost'** information now coming to light is surprising and has exciting implications. This is the first reasonably complete understanding of the sequence of events since one remarkable individual saw the need for such an Institute.

Celebrate now with us, read about the triumphs of the Institute and the many success stories of the members. Conversely, commiserate with us over the agony of the organisation when faced with a cruel turn of events.

THE NEED FOR THE INSTITUTE

Modern society appreciates the need for an organisation such as the IPI. This was not so in 1919, when the Institute was founded. A decade after its formation, the aims of the Institute were relatively similar to present day ones. But in 1919 there were major difficulties for inventors/patentees and licensees. These problems were peculiar to the times, a direct consequence of World War One.

It is well established that the founder was ***Mr Morgan Morgan, KC,*** a prominent London barrister. He was not an inventor, nor a Patent Attorney. Nevertheless, he became convinced that a strong organisation was needed to tackle these problems. After some months publicising his intentions to create an Institute, he arranged a public meeting which was well attended by potential members, mainly representatives of manufacturing companies and technical institutions. Many professional and business colleagues gave him their unqualified support. Individual inventors were not too evident initially, most of those joining were licensees, directors of companies with rights to manufacture and market new inventions. The men who attended the small private meeting to form the Institute were not necessarily the same people who subscribed to the first Memorandum and Articles of the company. So he was using his supporters in very specific ways. **MORGAN MORGAN** thus has every right to be spoken of as **THE FOUNDER** of the Institute. But many people gave tremendous support in early years. Some provided free office and secretarial facilities, others performed legal tasks at no cost or financed specific items. Within a few years the membership was high and then included individual inventors, whilst still maintaining a strong core of manufacturers and professional experts.

In order to understand why someone like Morgan chose to create the Institute, and why early members were licensee-oriented, an understanding of the economic climate and patent law of the time is needed. The victory in war was pyrrhic. The economy was stagnant, it was the age of industrial revolution but there was no money about to create new production lines. Inventors found manufacturers interested but unable to develop new products. Materials were unavailable or too costly. To make matters worse, manufacturers had contracted to produce new types of items, just prior to the war, but had been instructed by the government to make vital items for the war effort instead. Patent lifetime was ebbing away without the manufacturers being able to fulfil agreements, costs had been incurred but to no-one's profit. The UK Patent Law was not helpful, the USA, Germany and France had modernised their patent laws (giving longer patent lifetime, but the UK had not). So British manufacturers were at a disadvantage. Dominions of the Commonwealth had been in close collaboration throughout the war. This made a single patent application mechanism for the entire Commonwealth desirable in order to reduce costs, maximise product potential and give British industry an advantage in international markets.

It seems certain that Morgan Morgan had represented companies in conflict with the government over such post-war difficulties. He may also have been counselling individuals who had filed patent applications for inventions, but whose publications had been postponed/neglected (because of military implications). In many such cases, the product arising from the invention had been used in large numbers by British and allied forces during the war. After the war, the inventor had not even been offered a royalty let alone an account of the amount of usage. Invariably, the inventor had to sue the government for royalties and loss of patent lifetime. Cases often resulted in legal victory, but not great remuneration. Obviously, few individual inventors could afford to pursue this legal process. These cases did not involve dispute over ownership of intellectual property rights, it was ordinary law, not requiring a patent attorney. Many legal experts of that period were highly critical of the government's attitude to the rights of inventors and industry.

This then was the background to the need for the Institute, and why it was important to have many powerful legal, political, industrial and scientific professional people as members (rather than just inventors). Because the organisation was formed to take issue with the government, particularly The Patent Office, The Board of Trade, and The War Office/Defence Department. The times were desperate, individual or corporate inventors and licensees all had serious worries and complaints.

FORMATION OF THE ORGANISATION:

On exactly what date in 1919 the Institute was founded is a matter of opinion, the original intention had been to register a company called the **'The Imperial Institute of Patentees & Inventors'**, but this turned out to be unacceptable to The Board of Trade. Their first sentiments seem to have been for a name like **'The UK (or British) Patentees Protection Society'** but then, having in mind a Commonwealth-wide patent application mechanism, they opted for **'Imperial'**. A complete set of Memorandum & Articles was printed in 1919, which included references to unified Commonwealth patent mechanisms. However, when submitted to The Board of Trade at Companies House, the Memorandum was not to the government's liking and registration was refused because of the extensive legal implications. All Imperial & Commonwealth patent ambitions were then stymied. A copy of this rejected incorporation application is on file, a great deal of effort had gone into its preparation.

It is unclear exactly when this 'Imperial' registration application was submitted, it was at the Board of Trade offices for months while they mulled over the intentions of the new group. It must have been prepared from July 1919, it was rejected about June 1920. Consequently, the Institute was well into its second year before completing incorporation, the number of members was already several hundred. The first meeting on record was held by ***Morgan Morgan KC*** in his legal chambers, at 5 Pump Court in The Temple, London. This was an exploratory meeting attended by five individuals to discuss the concept of an organisation to protect the rights of inventors and patentees. Their primary concern was to pressure The Patent Office to extend the lifetime of patents that had been 'in limbo' because of the war. This was on May 10, 1919. It was agreed to arrange a public meeting and see what support there might be. The public meeting was on July 10, 1919, at the Cannon Street Hotel. ***Sir Joseph Lawrence, Bart,*** the owner of a company called 'Linotype and Machinery Ltd', chaired the meeting. Attendance was good, the motion to proceed was moved by ***Sir Herbert Nield, KC, MP. Sir Joseph Lawrence*** accepted the Inaugural Presidency, however this was a short-term appointment since his health was declining. In fact, Sir Joseph died in 1920, ***Sir Harry Foste***r took on the Presidency, also reluctantly. It was 1925 before a suitably prominent individual took the Presidency on for the long term, in the meantime there had been four 'interim' appointments.

Their problems with the proposed name and constitution had occupied an inordinate amount of time. However, another application, in the name of **'The Institute of Patentees'**, was approved and registration proceeded. This final application was received at the BOT on September 22, 1920, There was notification that a licence would be issued on September 27, 1920. The non-profit purposes were approved and permission was granted to use the term incorporated in the name, without using the word limited. The actual published registration documents were dated October 18, 1920. The first AGM was not held until January 12, 1921.

So on which date should we consider the Institute to have been founded? AGM and anniversary dates lead to confusion, some AGMs in the records are not recorded properly because it was forgotten that the first AGM was NOT held in 1920 (e.g., 46th AGM in records being actually the 45th). The incomplete archives have contributed to this confusion.

By September 1920, there was a substantial membership (170 members from the public inaugural meeting in July 1919, by the end of 1920 the number was around 700 and by the first AGM in Jan 1921 there were 1,000 members). Incorporation actually marked their first anniversary after realising by September 1919 that their original purposes were not to be. A practical way of treating this is to regard the end of 1919 as the foundation date, since it was by then clear what the organisation would be allowed to do, even if actual incorporation took a year to formalise.

THE ANNIVERSARY DATE:

The dinner arranged for November 1999 will therefore serve as the 80th anniversary of the Institute, as well as a social event to end the year.

INVENTORS HALL:

Getting suitable premises was difficult, given the range of activities desired in one Central London setting. It was on record at several of the first annual dinners that the Institute did not have its own premises, but operated out of members law offices. Finally in 1927, they obtained their own address, **39 Victoria Street, Westminster**, which was adequate for general purposes. However, in 1934, a benefactor who preferred to remain anonymous (a member who had benefited from the Institute's help) learned of premises that were ideal for all purposes. He/she donated a significant sum of money to enable the Institute to get the lease. The management moved rapidly and a month later the magazine illustrated the facilities - already in use. This was at **10 Victoria Street, Westminster**, the Institute name was prominent on an imposing frontage, there were ample offices, rooms for members to book for private meetings, a 200-seat lecture hall, a spacious library and reading room, and even space for light-industrial (non-noisy) experimental work as an aid to members and the scientific and technical committees. Within 3 years there was also an **'INVENTORS CORNER BAR'** with an additional annual subscription (but ladies not admitted). For the first time their AGMs were held in-house.

There are suggestions that the entire operation was referred to as **'INVENTORS HALL'**, it certainly had everything for the purpose. There had been references to the need for an Inventors Hall as early as 1930 anyway. The present management and council will be envious when looking at the scope of these pre-war facilities. Between 1921 and 1938, the membership had risen from 1,000 to 2,000, so the facilities needed to be comprehensive. The offices were open to members until 7pm on weekdays.

THE HOUSE MAGAZINE:

In 1938 the Institute magazine, ***'THE INVENTOR'***, which enjoyed considerable success on the news-stands (guaranteed circulation of 10,000 copies monthly), was modernised with a view to promoting sales worldwide. A professional publisher, ***Mr Herbert N. Casson***, was appointed **Editor**. A Managing Director was also appointed for the magazine's activities, ***Mr Edward O. Beck***, who had worked with Mr Casson for many years. The intention was to increase sales to at least 50,000 per issue. Advertising and editorial activities were run from Mr Casson's offices, and business accounts were in the name of **'Inventator Ltd'**.

Herbert Casson was no stranger to the world of invention, he had written up the works of Lord Kelvin, Alexander Graham Bell and Cyrus Hall McCormick. Scores of other inventors had benefited from his help, including the Wright Brothers. He advised them about patents, then wrote the first published story about their flight achievements. So the Institute had the right team for this publishing expansion. Immediately, the new series of journals became technically sophisticated, comprehensive and excellent value for money (30-50pps at 6d/copy).

All good things do come to an end, and due to the international situation, in August 1938 the **cancellation of the 1939 exhibition** of inventions was announced. The Institute journal ***'THE INVENTOR'*** ceased publication after July 1939, there were **no more journals** until 1948. It was 1961 before the next professional-looking publication appeared, and that was quarterly, not monthly.

FRIDAY THE THIRTEENTH:

The cruellest blow of all was when these ideal premises were destroyed by fire. On Friday, September 13, 1940, at 2200 hours a high-explosive bomb hit No.14 Victoria Street and adjacent buildings were set on fire. Apparently 3 people died in the event (probably not from the Institute). Rochester Row police station recorded it, but it was 2am next day before emergency services had all responded, they were no doubt attending critical blazes already. **The date, the 13th, not to mention being the Friday!**

Undoubtedly, loss of records and technical items was severe. The organisation's next residence was at **50 Pall Mall**, which seems to have been the head offices of ***Sir Arrol Moir***, a well-known and successful chemical engineer/trader (a member). No doubt remaining records were dispersed to homes of staff and council to minimise future space needs. As always, material got lost in the shuffle.

Until this point **(here endeth part one)**, the Institute had been a huge success and later it will be seen that original concerns had been reasonably well resolved. But who were the people responsible for getting the Institute this far, and at such speed?

THE FOUNDER MEMBERS:

At incorporation, in 1920, there were **7 Subscribers** and **7 Council Members** detailed. **Morgan Morgan** and ***Herbert Ernest Dakin*** (an inventor), were listed as **Subscribers** and **Council Members**. The other **Subscribers** were:-

> ***Sir Harry Seymour Foster, JP, DL, MP (Consul General of Persia).***
> ***Sir William Grey-Wilson, KCMG, CMG, KBE (Ex-Governor).***
> ***Mr Joseph Ponsonby Horatio Bewsher (Journalist).***
> ***Mr Godfrey N. R. Nightingale Cheeseman (Company Director).***
> ***Mr Arthur Ernest Lake (Engineer).***

The other initial **<u>Council Members</u>** identified were:-

> ***Mr Ion Darby (of Weston Metal Works).***
> ***Captain E. Thomas Elias (Shipowner).***
> ***Mr A. G. Hardcastle (Chairman, Madgwick Ltd).***
> ***Mr Samuel Matthew Johnson (Engineer).***
> ***Mr George Herbert Skinner (Managing Director, S .U. Company Ltd).***

These individuals took on the legal and financial responsibilities. This **Executive Council** managed the Institute until the **first AGM on January 12, 1921**. The number of Council Members was increased to 21 at the first elections, and this number was maintained for many years. The Founder Council Members were permanent. Several committees were created (legal, financial, commercial, political, technical & scientific) and each Council Member played a part on one or more committees. All Officers of the Council occupied honorary positions. Little is known of the roles of these few Institute Founders, except that ***Morgan Morgan*** functioned as **President** temporarily one year, and as **Secretary** in another year. ***Joshua Bewsher*** was **Treasurer** for the first few years. ***George Skinner*** (inventor of the S. U. Carburettor) was still on the Council at the time of his death in 1933. ***Sir William Grey-Wilson*** was well known to the public, having been a Governor of various colonies. At the time of death he was a **past President** and current **Chairman** of the Institute. ***Sir Harry Foster*** had been the **Sheriff** and **H. M. Lieutenant of the City of London** (hence the knighthood). All of these people were prominent in the City.

The organisation had been living dangerously by operating whilst not registered, some of the legal/financial committee were worried about this.

Sir Joseph Lawrence, the first **President**, died in 1921, ***Sir Harry Foster*** took over the role prior to this and was then elected to the position for 1921-2. **The Secretary** was ***Captain G. Drury Coleman. Professor Archibald Montgomery Low*** joined the Council in 1921 giving more depth to scientific appraisals. The second AGM was held on 10/3/1922, the third on 21/3/1923, but there is little more information available until 1925.

QUOTES FROM THE FOUNDERS:
The following comments from Institute management throw some light on the difficulties they faced.

> "Government and the public classify patentees along with clergymen and widows, i.e., as unbusinesslike".
> "Licensees were paying taxes, but not to personal gain, with some patents lapsing at critical stages".

"Government takes something for nothing, by accepting fees for patent applications they know at the outset are not remotely new - i.e., unpatentable".
"Not one UK inventor had got rich, this is wrong".
"Still making empire-oriented speeches, particularly at The Royal Colonial Institute".
"Would someone please invent me something to say when I come home late at night (to the wife)".
Frequent references to the 1907 Patent Act, which had reduced patent lifetime from 40 years to 16 years, for practical purposes this was considered a mere 14 years of useful life.

THE A-TEAM:

By 1925, a powerful, efficient and long-term management group was developing. ***Lord Askwith*** had been elected **President** and remained in office until 1942. ***Captain Coleman*** was now well settled in as **Secretary** (and became the driving force behind the annual exhibitions), ***Professor Low*** had reorganised the **technical/scientific** appraisal group. ***Mr S. E. Denning***, elected as **Treasurer** in 1925, served until 1954. Within a year, **Professor P. M. Baker** was brought onto the Council, and the technical & scientific committee split into two groups (one chaired by Low, the other by Baker). The Professors were the **first Fellows** elected (1926).

PROFILES
(cast in order of appearance)

Mr MORGAN MORGAN, KCB. Died on August 29, 1930.
Born of a poor Welsh mining family, and worked for a short while in the pits as a youngster. He moved to Cardiff and bettered himself. Twice he was a Labour/Liberal representative on the City Council, then Deputy Mayor of Cardiff at age 29. A JP a little later, then into Law seriously and became a barrister in London, KC. Morgan was planning to enter politics, but on becoming deaf he changed his plans. His venture with the Institute formation was part of necessary diversification. Highly energetic and a pillar of strength for the Institute, undoubtedly a powerful speaker but reluctant to be President, partly because of his hearing difficulties.

SIR HARRY SEYMOUR FOSTER, JP, DL, MP.
A Founder Member. A Chartered Accountant until 1886, then Sheriff of London in 1891 followed by 40 years in Parliament. The Consul General for Persia from 1894-1923. Knighted in 1918, following duty as H. M. Lieutenant of London.

SIR WILLIAM GREY-WILSON, KCMG, CMG, KBE. Died in 1926.
A Founder and strong supporter in the first years. After a career in the Colonial Office in Jamaica and elsewhere, he became Governor of St. Helena, The Falkland

Isles and eventually The Bahamas. When he died he had been a Past President of the Institute and was Chairman at the time. He was a grandson of Earl Grey.

CAPTAIN GEORGE DRURY COLEMAN, FCIS. Died in 1968.

Educated at Sherborne, and then Imperial College/School of Mines, worked for 'The Times' before joining the Institute. General Secretary from 1920 until 1940, when he joined his Regiment (Norfolk). Retired to Southwold after the war. He had served in both wars. Considered very good company. He masterminded the international exhibitions and competitions efficiently.

PROFESSOR ARCHIBALD MONTGOMERY LOW, DSc, ACGI, FFSc, FICA, MIME, MIAE, FCS, etc. Died in 1956.

A member of the Council and Scientific Committee/Honorary Consultant to the Institute from 1921, A Fellow from 1926, Vice President, then President from 1942-56. A prolific inventor, writer and public speaker. Particular interests were space-travel, clairvoyancy, Acoustics/noise-measurement, weapons and explosives. Hon. Assistant Professor of Physics at The Royal Artillery College, past Chairman of The Autocycle Union and of The British Auto Racing Club, also ex-President of The Interplanetary Travel Society. With Lt. Col Moore-Brabazon (later Lord Brabazon) he produced and edited ***'ARMCHAIR SCIENCE'*** written for laymen. An inventive period was in WW1, when he was Commandant of The Royal Flying Corps experimental works. He demonstrated the first **'flying robot'** (radio-controlled rocket), created several explosives, improved radio-signals apparatus and certain weapons (machine-gun synchronisation, flying bomb relays). In WW2, he served in The War Office on improving explosives. In 1914 he demonstrated the principles of television. An expert on **noise-reduction** in planes/ underground-trains/public buildings, he devised the **low-Hilger audiometer** which photographs sound, a much used device. His death was a massive blow to the Institute, since he had been a workaholic who knew the invention business from all angles. He was not a one-man Institute, but chaired almost all meetings and wrote articles for every magazine. When he died on Friday September 13, 1956, no-one was ready to try and step into his shoes. In fact the Institute had to be run for 5 years by the Chairman.

Rt. Hon. The LORD ASKWITH, KCB, DCL. Died in 1942.

George Ranken Askwith, born in 1861, he took History at Oxford, then studied to become a barrister and practised from 1886, KC obtained in 1908. His career then concentrated on arbitration/conciliation, he was a notable Comptroller of the Commercial, Labour & Statistical Dept, Board of Trade from 1909-11. Then Chief Industrial Commissioner until 1919. A KCB in 1911, and made a Baron in 1919. President of the Institute from 1925-42. A powerful force for the organisation because of his standing with the government. A relevant previous appointment was that of Comptroller of the Patent Office. Known as **'Askwith the Peacemaker'**.

Mr S. E. DENNING, FCIS, FSS. Died in 1955.

A Member of the Institute since about 1921, and Hon. Treasurer from 1925 until death. Vice President from 1951. His Accountancy training had been with a prominent Patent Attorney, then he was Executor for the will of Lord Grimthorpe, a task involving much innovation material/assets. So he was excellent for the Institute's needs. One profile of him said he never has any temptation to 'open the cigar box'. He was a lover of good order, and his financial skills enabled the Institute to survive in the difficult times after loss of the premises in Victoria Street, in 1940.

PROFESSOR P. M. BAKER, MBE, BSc, AMIEE, AMIMechE, FInstA.

Born at Aston Manor in The Midlands, in 1872. During 7 years teaching at Wolverhampton Technical School, he developed an outstanding Dept of Engineering. Then he became Professor of Engineering & Physics at The Central Technical College in Bombay. There he succeeded in instilling a mechanical mentality, which was not natural to Asian students at that time. During WW1, he was Professor of Munitions and Director of Training for munitions workers (UK), his systems became commonplace in factories. After the armistice he joined The Ministry of Labour. An Hon. Fellow of the Institute from 1926, Chairman of one of the technical committees, and on the Scholarship Committee. His loss only 4 years after reorganisation of the way the Technical & Scientific Committees operated, was most unfortunate for the Institute. Died in 1930.

ACHIEVEMENTS

Clearly there were 3 sets of aims. To press the government for improvements to the current Patent Act (1907), to get financial rewards for applicants and licensees affected by the wartime situation on patents, and provision of more services to current inventors/members. What success did they have in these respects?

A. PATENT LAWS. Eventually the Patent Law was changed to provide a longer period for examination of commercial viability prior to full processing of the application. This was lengthened first to 9 months and then to 1 year (1919 Act). Later the Patent Lifetime was extended, from 14 to 16 years, and then gave a protection period of 20 years as at present. So the Institute had fulfilled an important role in these areas by defining exactly what was necessary in the way of change. But it was 1938 before The Patent Act was entirely modernised.

B. WARTIME ROYALTIES. On the other front, a War Commission was set up by the government, to examine claims by aggrieved inventors whose new products had been used in the warfare. The Commission consisted of 7 legal experts (unpaid), and it held over 400 sessions between 1919 and 1933 examining claims for financial compensation. The report issued in 1933 indicated that some 1,850 claims were submitted, many were unsubstantiated and rapidly discarded. Retroactive payments made to about 450 individual inventors and companies totalled £1,500,000. Individual awards were amicably received (with a few notable exceptions). Apparently, 70 people had claimed to invent the **barrage balloon system** as used in London's air defence. Large awards were for aircraft engine designs, £159,000 for the 'Gnome' and 'Le Rhone' engines, £65,000 for De

Havilland planes, and £50,000 for Bristol fighters. Another large award was to Messrs Peter Brotherhood Ltd (£125,000) for improvements in internal combustion engines, **but the details were still kept secret!**

Another area of confusion was determining which of the 85 claims to be **the inventor of the tank** were most relevant. ***Sir William Tritton*** and a ***Major Wilson*** each received £15,000 for their patented tank designs.

The claims covered a vast field, tanks, aero-engines, improved guns, warships, submarines, torpedoes, bombs and shells. Other interesting devices included smoke-screen production, anti-dazzle camouflage painting of ships, kite-balloons, every type of armament and transport by land/sea/air was presented to the Commission. However, many claimants could not meet the essential condition proving - **'a link in the chain of causation leading to the use of the invention**'.

Mr Martin F. Hale, inventor of rifle grenades and aerial bombs as used extensively in the war, was awarded only £22,000, whilst his legal costs were £13,000 and tax/supertax took another £8,000 of the award. He claimed that any ordinary business user of his products would have paid out £300,000 in royalties for that amount of usage of these inventions. He was not alone in feeling mistreated by the government. An additional irritation for him, which he publicised widely, was that the taxman got his cut before he had got his money. In his last public comments, he claimed that the final reckoning meant he had been fined £14,000 for helping his country.

C. MEMBER SERVICES. In regard to general services for members, an amazing variety of facilities were on offer. A member could apply on a dozen special forms for services concerning patent applications, searches, royalty collection, product development, promotion, licensing, distribution, you name it. But irrespective of a particular member's needs at a given moment, there were many general benefits. The Institute logo could be used on one's personal stationery, an annual diary and pocket book with details of current Patent Law and application procedures were also available. The magazine had prize competitions for exhibition poster designs, new logo competitions were profitable to some, there were prizes for the best **'What's Wanted'** suggestions at the exhibitions. At the London offices, one could meet in private with potential industrial partners, or use the technical library/patent collection, make a prototype or just sit around the 'Inventors Bar' talking about it. The office was even available as an accommodation address for individuals or company members. Any amount of technical and legal advice was free to members, plus you could get your **tax return** sorted out there.

There were frequent meetings, and from 1925 **annual exhibitions** of inventions (held at Central Hall, Westminster). During the 30s when the Institute was expanding fast, several major inventions exhibitions were held in the North as well.

These were successful (usually in Newcastle or Manchester). 1937 was unusual in that exhibitions were held in Leeds and Sheffield but not in London. 1934 was notable in that **The Prince of Wales** (later King Edward VIII) visited the 10th international exhibition. One Vice President, ***Sir John McDonald, JP***, from Glasgow, was so impressed by the visit occurring that he donated £250 to Institute funds.

In the 1930s, there had been **European Invention Exhibitions** held in **Brussels** until the war. These were well attended, and it was the practice for British inventors to enter these competitions via the Institute, which acted as their UK offices. The British contingent was always strong and they carried off many of the prizes. In one year 15 British entrants won awards, most were of course members of the Institute.

The '**What's Wanted'** columns and books had an unusual origin. In an editorial in the first magazine (October 1929), ***The Rt. Hon. Sir William Bull, Bt, FSA***, admits to having carried a list of wanted inventions around with him for several years. He presented the manuscript of the little book to the Institute and it was published and upgraded every few years. His interest in these commercial prospects stemmed from being Chairman of J. W. Singer & Sons, and a Vice Chairman of The BBC and Siemens Bros. Ultimately the subject was dropped because of inevitable redundancy in the material. Members were requesting items that unknown to them were already on sale somewhere in the World (usually in the USA). Sir William died in 1931.

Throughout most of the Institute's history there have been different membership classes. Ordinary members, maker members, associates, fellows, student categories, honorary members/fellows/vice presidents, life-members, although the terminology changed periodically. The relative significance and manner of gaining a higher status/grade has also evolved.

AFTERMATH OF THE DESTRUCTION/GETTING STARTED AGAIN:

Part two of the life-cycle concerns the years immediately following the loss of 10 Victoria Street, due to the bombing, and then up to 1974-5. This difficult period starts with the financial implications of beginning again in someone else's premises. The Institute survived this unpleasant shock, despite it happening when members were away at war and funds were low. But in time they got situated properly, in Victoria Street, again, adjacent to their original premises (rebuilt). Researching the Institute's whereabouts was complicated by the fact that it had occupied **Nos. 2/8, 10, 25 and 39 Victoria Street**, at various times. Most of these addresses seem to have been called **Abbey House** at one time or another.

The existence of undated documents that refer to office numbers within Abbey House, but do not give the building number, was driving the researcher up the wall for a while! And there is no correlation with present day buildings since the area has been redeveloped several times, these buildings are now a major police HQ.

In any event, it is now evident that the 50 Pall Mall address was used until 1946. Then it was back to **Victoria Street**, again **(No.25, South Block)** for 2 years (this may not have been their own offices either). Finally, in 1948, the Institute gained a 25-year lease at **2/8 Victoria Street, Nos. 207/8 Abbey House**. They had reasonable space but only offices. The end of this period was again marked by a need for fresh premises, with little warning the building on Victoria Street was earmarked for redevelopment in 1974. The Institute had anticipated a renewal of lease, instead of which there was a frantic search for a new place 2 months before the lease-end.

In 1948, there were declarations that the magazine 'THE INVENTOR' would be published again soon. However, there was a single issue of 'NEWS REVIEW' in February 1948, with an immediate change of name to '***INVENTORS NEWS***'. This quarterly issue ceased abruptly in April 1957. When reactivated in 1959 as the '***IPI NEWSLETTER***' it was just a set of news-sheets. But in March 1961, the first real issue of ***'THE INVENTOR'*** since 1939 was introduced, still a quarterly, but colour-coded issues with topical cover photo's and professional appearance. The title theme for the new series was **'The only magazine for inventions and inventors'**. The March 1969 edition was a 50th anniversary issue and had 70 pages. In addition to information about the exhibition and dinner events, it contained transcripts of papers presented at the Conference. These presentations concerned various modern research topics, including protection of industrial and intellectual property.

A surprising item found in a 1948 copy of the newsletter was a request from the Institute for suggestions as to what might be the best criteria to include in an '**INVENTORS CHARTER**', a charter for the government to observe. It all read very much like 1999 people's inventor stuff! This might relate to the fact that in 1948, ***Sir Arrol Moir*** and ***Professor Low*** visited the Patent Office in April, during a sitting of the **Royal Commission** considering various revisions to the **Patents Acts**. They detailed the views and complaints of members of the Institute.

Two important changes were made to the organisation in the 50s. In the first, at an extraordinary general meeting on January 21, 1954, another class of membership was adopted. This was '**Companion**' status for interested young people, permitting reduced fees and some rights and privileges but not a vote. There has usually been provision for students but this catered for teenagers.

The next special meeting, chaired by Professor Low, and again at the Central Hall, Westminster, was on March 23, 1956. This was to change the name to **'THE INSTITUTE of PATENTEES and INVENTORS'**. Researching this led to some confusion because at the inaugural meeting there had been discussion about the existence of a Manchester-based group called **'The Institute of Inventors'** and should they not merge from the start. This suggestion was not followed up. But apparently this other group, mainly of inventors (rather than licensees), was several hundred strong and had been formed in 1913. ***Mr Walter Reid*** was their Chairman (the inventor of smokeless gunpowder). Mr Reid and others were at the meeting and no doubt then belonged to both organisations. The Northern group dispersed during WW2.

However, it did become clear that this change of name merely reflected a growing recognition that the members were now mainly inventors not licensees and it was preferable to have a name that properly represented the Institute's purposes and structure. The new name had been selected by the management and presented to the meeting, a few members present felt the management should have consulted them on choice of name, and one suggestion from the floor was for a name like **'The Society of Inventors'**. But it was explained the Institute had already taken legal advice on this and names such as that one were already in use anyway, the change proceeded without further debate.

There were a number of changes in the **Secretary** office during this period, ***Sir Arrol Moir*** appears to have filled that role from 1941-1948, presumably as a consequence of the Institute operating out of his offices. Then ***Major Frederick Charles Beauchamp*** held the office for 6 years, then ***Major R. O. Sutherland*** from 1954-60, followed by ***Mr Alfred Leonard Tams Cotterell, MBE, MIPR***, who was Secretary from 1964-78. ***Mrs Hazeldine*** was Acting Secretary in some of the interim period 1960-64, as was ***Lt. Col. Ambrose Crawley***.

However, the most significant event of this period was the untimely death of Professor Low in 1956. The date was **September 13th** (again), but it was NOT a Friday (an hour off since it was a Thursday). He had been President from 1942, after Lord Askwith. Because he had been such a 'hands-on' leader, his death paralysed the Institute for almost a year. He had been the first President who was a highly qualified scientist and inventor, marking a change from previous presidential practice. The Council hardly knew in which direction to look next.

It is now evident that the period 1956-1961 was without even temporary Presidents. ***Arthur W. Richardson*** may have been the **Chairman** prior to Professor Low's death or not, records are not available. However from 1956, he managed the IPI absolutely, until 1962 when ***Sir Stanley Rawson*** accepted the Presidency. The most important feature of the period is that it marked a distinct and permanent upgrading of the significance of the Council Chairman in the overall control of the Institute. Future Presidents were not merely figureheads, ***Sir John Maitland*** and ***Lord Lloyd*** being highly effective in raising the status and profile of the Institute and inventors. But the Chairmanship became much more publicly visible as the management.

A deputation from the Institute met with the Patent Comptroller in 1961, to protest at the long delays before grant of a patent. ***Mr Sam Green*** and ***Mr Arnold Tickner*** (a Patent Agent and Member) pointed out the hardships inventors suffered as a result of the time lag.

THE INTERNATIONAL SCENE

In the early 60s a lot of communication was established between inventors groups in various countries. In particular, the UK and Nordic countries were very active. Following consultations between ***Mr Harald Romanus***, representing groups in Finland, Sweden, Norway and Denmark, and the Council of the IPI, an international organisation was created. This was named the **International Federation of Inventors Associations (IFIA)**.

The first meeting was held in London in 1968, the IPI provided the **Secretariat** then, and the **first President** was from the IPI, ***Mr Arthur W. Richardson***. The **World Intellectual Property Organisation (WIPO)** accorded IFIA status as an observer, which meant that this non-governmental body had some input into the policies.

Institute officials who have played a part in the IFIA function includes ***Mr Sam Green*** who became IFIA's first **Chairman** in 1968, and who as Deputy Chairman of the IPI originated the talks with Harald Romanus. Initially ***Mr A. Leonard Cotterell*** was the **Secretary**, many years later ***Mr Pat Russell-Rayner*** took on the position. ***Sir John Maitland***, the IPI President oversaw the inaugural meetings. So some of the most experienced IPI officials were concerned with the initiation, and therefore a reasonable degree of success was guaranteed.

Shortly before gaining his peerage, ***Mr G. Lloyd, CBE, QC, JP, MA, BSc***, a former barrister, represented the IPI at an **IFIA Diplomatic Conference** in Munich (1972). An interesting aside is that it was reported in the journal that when taking up his peerage he would no doubt become ***Lord Gerran Lloyd***, but in fact he opted for ***Lord Lloyd of Kilgerran***.

THE 50th ANNIVERSARY (1969)

In 1969 the Institute sponsored The London International Inventions and New Products Exhibition **(LINPEX)**, held at The Royal Horticultural Hall on January 6-11. This was undoubtedly the major single undertaking by the Institute up to that time. The exhibition was opened by ***Mr A. G. Norman, CBE***, President of the CBI. Over 200 exhibits were on show, and there was a distinct international flavour in the exhibits and some groups of visitors. More than 7,000 people attended the exhibition. Concurrent with this was a conference held at the Royal Horticultural Society in London. 256 delegates attended this conference at which 8 major papers were delivered on the theme of 'The importance of patents and inventions and their exploitation by industry'.

The associated **Anniversary Dinner** was held at The Cafe Royal, London, on January 10th. The Guest of Honour was the then Minister of Technology, ***The Right Hon. Anthony Wedgwood Benn, MP***, who during the programme awarded the prizes and trophies to the outstanding exhibitors. The panel of judges included ***Arthur W. Richardson*** (Chairman of IPI), ***Sam Green*** (Chairman of IFIA), ***Mr O. W. Lipton*** (Vice President). ***David Fishlock*** of the 'Financial Times' and ***Peter Dew, OBE***, of Westminster City Council made up the panel.

In 1972 an **Extraordinary General Meeting** was held. The Articles and Memorandum of incorporation were entirely modernised. All references to Dominions and Empire were deleted, decimalisation of sterling was recognised in reference to fees, and modern legal terminology replaced obsolete wordings.

WARTIME INVENTIONS COMMISSION (WW2)

In 1973 the journal announced that ***Lord Lloyd of Kilgerran***, an Honorary Vice president, had been the **Secretary** to the **Wartimes Inventions Commission** and had arranged for records to be made available for archival purposes. However, there is no evidence of published details, and according to ***Mr Pat Russell-Rayner***, the document transfer was intended but not fulfilled. It is a pity because these WW2 royalties arguments would have made interesting reading, as did the WW1 details aforementioned. Albeit retroactively this might be good material for a future article.

NEW PREMISES: Moving premises just before Xmas in 1973, was a diabolical business. It was the time of rail strikes, go-slows, the 3-day week, the removal firm was something else, staff couldn't get to work, and the phone company let them down. It was apparently not much fun. The new address was:

The Whiteley Building at 165 Queensway, in Bayswater W2.

The result was a trade-off at similar cost to the last premises, The location was accepted as not that convenient for members, but they had more office space. There might have been some other benefits in that **The Science Reference Library** and ICL (computers) were occupants of the same building. These offices were used for 5 years.

Another difficulty with the removal time was that management were busy organising a **Reception at the House of Lords** for early in 1974. Hardly done under ideal circumstances, nevertheless the occasion was successful.
A member had donated £1,000 to the Institute to fund a reception that might aid the process of educating more politicians to the world of invention.

It was held on January 15, 1974, in the Chelmondeley Room at the House of Lords, with ***Lord and Lady Inglewood*** as the hosts (he was an Honorary Vice President of the IPI). The 60 guests included many parliamentarians and industrial leaders. The President of the Institute ***Sir John Maitland***, and officers of the IPI attended. Vice Presidents, ***Mr Arthur W. Richardson, Mr Sam Green, Mr L. W. Gordon Davies***, and ***Mr O. W. Lipton***, plus Council Members ***Mr Bill de May*** and ***Mr W. O. Nutt***, the Secretary ***Mr A. Leonard Cotterel***l, and Assistant Secretary ***Mrs da Silva***, were all present. Guests associated with the Institute were ***Sir Christopher Cockerell, Mr Richard Luce, MP, Lord Ironside*** and ***Lord Lloyd of Kilgerran***. The Minister for Trade and Consumer Affairs then, ***Sir Geoffrey Howe, MP***, was the principal **Guest of Honour** and addressed the meeting.

There were in fact 2 Guests of Honour on this occasion, since ***Lady Legge-Bourke*** was present and the occasion also served as a tribute to her late husband, who had served the Institute remarkably well during his Vice Presidency.

Not often referred to in print was that ***Sir John Maitland, Lord Inglewood*** and ***Lord Ironside*** considered themselves to be the **'Three Musketeers'** of the world of invention, battling to get the government to view inventors in a better light. They did in fact achieve a better status and more acceptability for the inventor, during their terms of office. Not seen put that way, but a conspicuous parallel, is that ***Cockerell, Tedder*** and ***Wallis*** were like 3 Musketeers too but on a more individual plane.

Periodically there has been some fast and furious correspondence on unusual topics that attracted a lot of interest, examples have been:

A monkey climbing up a rope, scientific speculation.
Perpetual motion, a particular clock invented by a Frenchman
(operated by changes in the external atmosphere).
The laws of **vibration**.
New theories of **gravity**.
The age that inventors tend to be, when most successful, and which astrological group they tend to be.

A more serious topic that occurred regularly in correspondence and articles in this period, was the role and rights of **employee-inventors**. This debate was global, not confined to the UK, IFIA bulletins reported on how different countries coped.

Introduced during this period were **crosswords** for inventors, **quizzes and puzzle pictures** (a 'thingy' photographed from an unusual angle - what was it?).

As already mentioned, the office addresses all on Victoria Street were hard to sort out. A less complex item but more bizarre was the inclusion of ***Professor Lord Tedder*** in the panel of judges at LINPEX 69, as listed in the special edition of the anniversary magazine published later. His wife ***The Hon. Mina Tedder***, was listed as a guest at the President's table at the dinner, but Lord Tedder was not listed. The entry read Professor Lord Tedder, MA, PhD, DSc, ScD, Professor of Chemistry, Dundee University. He was a judge along with Richardson, Green and Lipton. But according to other information already compiled, Arthur Tedder died in 1967. It must have caused some embarrassment at the time of publication. Nowadays we would refer to it as bad work by a **ghost writer** (Pun intended).

PROFILES (post-war/mainly 1948-1974/5)

PROFESSOR A. M. LOW. Died in 1956.
Described in detail earlier. President from 1942-1956, he took office shortly before the death of Lord Askwith.

Mr S. E. DENNING, FCIS, FSS. Died in 1955.
Described earlier, one of the earliest and long-serving members, Treasurer from 1925 until 1955. A Vice President from 1951. His personal integrity was a major factor in the Institute surviving the loss of the offices in 1940 and subsequent revival.

CAPTAIN SIR ARROL MOIR, Bart, BA, MInstCE. Died in 1957.
Knighthood in 1916, an officer in the Royal Engineers in WW1 and WW2. Following Captain Coleman's recall to his Regiment (The Norfolks) in 1940, Sir Arrol became the general Secretary, and then in September when the offices were bombed, he provided the office facilities for some years. In 1948 upon retiring from the Secretary post he was made a Vice President. He was a bit of an artist too, since in 1936 he won the Poster Design competition for the next International Exhibition of Inventions in London.

MAJOR FREDERICK CHARLES BEAUCHAMP. Died in 1954.
General Secretary from 1948-1954.

MAJOR R. O. SUTHERLAND.
General Secretary from 1954-1960.

Mr ARTHUR WHITTAKER RICHARDSON, OBE, FInstPet, FCS, FRSA, Hon. FInstPI. Died in 1985, aged 91.
Chairman of the Institute for about 20 years, an Honorary Fellow and a Vice President. In the Petrochemical Industry, he was Manager of the Kerosene, Liquid Gas Development Laboratory for Shell. Winner of the **Diesel Medal** in early years. He invented an **unspillable domestic oil heater**, manufactured by a company he set up. He instituted the **Richardson Annual Gold Medal Fund**. The first **President of IFIA**. He may have been Chairman at the time of Professor Low's death in 1956. In any event he ran the Institute from then until 1962, without a President.

Mr E. H. C. WEBB, FIAC.
Treasurer from 1956-1968, the son of Mr F. H. Webb, a Council member.

Mr F. H. WEBB, LRIBA. Died in 1956.
An architect who served on the finance committee and was involved in the search for ideal multi-purpose premises in the 30s (prior to the Victoria Street offices availability). He died shortly after his son became a Council Member.

Lt. Col. AMBROSE JOHN CRAWLEY. Died in 1980.
A Member since 1955 and on the Council for many years. Being an expert on Industrial Law he served on the Legal/Patent committee. He was **Acting Secretary** for a couple of periods. He adopted the title of **'Technical Advisor to the IPI'** to emphasise when he was not speaking as a professional legal representative. His mother had once been Florence Nightingale's secretary. He was Royal Artillery staff in armaments inspection. Retired and went to The Bar in 1952. In 1944 he needed help with an invention, joined the IPI and then was on the Council from 1957.

Mrs M. D. HAZELDINE.
The longest serving employee, head of the office for many years and in the early 60s was Acting Secretary of the Institute. In 1963 she became the official Company Secretary. She made several attempts to retire but had served about 45 years before getting away. Awarded an Honorary Life Membership in latter years.

SIR STANLEY RAWSON. Died in 1973.
He accepted the Presidency in 1962, several years after Professor Low's death, and served until 1966. Probably not intending to do a long term of office. But he gave Mr Richardson, the Chairman, much help until Sir John Maitland took over in 1966.

COMMANDER SIR JOHN MAITLAND, RN (Ret), JP, FRSA. Died in 1977.
President of the Institute from 1966-1975. Recalled to the Royal Navy for WW2, then from 1945 an MP. He joined the IPI after enquiring about an invention for submariners. In 1966, he raised the question of support for the IPI in the House of Commons. He arranged a reception for IPI Council Members to meet with MPs, later the BOT gave some financial support. He presided over **LINPEX 69** and was crucial to getting talks about IFIA off the ground. Knighted in 1969.

ARTHUR T. A. NEGUS, BA, ACA. Died in 1977.
A Member from 1968, Treasurer from a year later and until death. A Chartered Accountant, a Cambridge graduate.

This marks the end of the description of phase 2 of the IPI history.

'NEW PATENTS' was a special publication produced by the Institute from the 70s through to 1995. It played a useful role in alerting industry to significant patents published in the UK. It was not for the ordinary member, a separate subscription was involved and IPI Company Members received it several weeks prior to non-Member subscribers. Items expected to be of interest to a particular company were highlighted. As electronic patent searches via the Internet became available there was little further purpose to it.

IFIA

The organisation holds its annual meeting in a different member country each year, it was held in the UK in the initiation year 1968 and then in 1975 and 1982. The present composition of IFIA is a mixture of about 80 Inventors' Associations, some Women Inventors' Associations, a number of affiliated organisations and many corresponding individuals and collaborating institutions in a total of 91 countries.

The **Secretariat** is now based outside the UK. In the 70s it was a valuable organisation which contributed to the formulation of the **EPC** and **PCT** patent systems. Much of its news concerned these, and the employee-inventor rewards controversy. These topics are no longer burning issues, the patent systems are in operation and work well enough. So the IFIA organisation played a useful role and is still worthwhile, but not quite as vital as in the 70s and 80s.

NEW PREMISES: in mid-1978 the Institute moved to offices at **Staple Inn Buildings (South Block)**, 335 High Holborn, on the 4th floor. This was a pleasant location in the legal environment and close to the Patent Office and Science Reference library. An elegant picture in the illustrations section depicts this location. This address was used until 1986, when the Institute transferred to its present address in **Regent Street**.

60th ANNIVERSARY MEETING was not on quite as grand a scale as the 50th, but it was notable, and the special edition of the magazine later had a nice colour supplement as an insert (which the 50th did not have). The meeting was held at The Royal Society of Arts, on September 19, 1979. About 30 members exhibited.

The **Principal Guest** was ***Mr David Mitchell, MP***, Parliamentary Secretary to Sir Keith Joseph in the DTI. The Institute's President ***Lord Lloyd of Kilgerran*** was present with Lady Lloyd. Other guests included ***The Lord Nathan*** and ***Dame Diana Reader Harris, DBE, BA, FCP***, who represented the RSA. ***Mr I. J. G. Davies***, Comptroller General of the Patent Office, and ***Mr M. W. Makinson***, MD of the NRDC, were also present.

'***FUTURE AND THE INVENTOR***' became the new name for the journal in 1987.

Dr Parsons, MA, PhD, an Honorary Member of the Institute, was working for the DTI, in the 1990s when he pointed out to his department the nature of the Public service performed by the Institute. His action was responsible for a grant being forthcoming from the DTI for several years.

CHARITABLE STATUS: At an Extraordinary General Meeting in 1995 it was agreed to apply for Charitable Status for the IPI. This would allow covenanted subscriptions and have positive VAT implications. Subscriptions would rise and VAT status would result in better cashflow. First reactions appeared favourable but the final outcome in 1997 was negative. An appeal was also unsuccessful. This decision of the Charities Commission was a severe setback. It remains to be seen whether another application will be prepared in the future.

At the same Extraordinary General Meeting, it was decided to undergo a complete **revision of the Memorandum & Articles of the IPI**. This preparation work also took 2 years. At the AGM in 1997, the improved Memo & Articles were examined and adopted, sent for reprinting.

The **NATIONAL INNOVATION CENTRE** was a project initiated by the **IPDC** (Intellectual Property Development Confederation) in 1995. It involved a proposal to the Millennium Commission to partly fund the building, and has been publicised in the *'INVENTORS WORLD'*. This national facility would cost about £10 million to create, an architectural design has already been selected. There is a **Steering Committee**, consisting of ***Dr Barrie Blake-Coleman (Chairman), Peter Knight*** and ***David Wardell*** (of **'Inventors World'**, and the IPDC then), ***Trevor Baylis, Georgina Richards*** **(Intellectual Property Solicitor)**, ***Paul Ambridge*** **(IPI)** and ***Ted Blake*** **(formerly of the Patent Office, now with CIPA)**. The concept has gained some useful support, but has yet to receive any funding on the necessary scale.

In 1997 the first Inventors Courses were made available, through the **Inventors Workshop** set up at Paddington College.

PROFILES (part three 1975-1998)

The LORD LLOYD of KILGERRAN, CBE, KC, MA, BSc. Died in 1990.
President of the Institute from 1975 to 1990. He had been a Honorary Vice President for a short period before Sir John Maitland died, when he then accepted the Presidency. Rhys Gerran Lloyd was educated as a scientist and as a barrister. He was an expert on Patents and Trade Mark Law. From 1971-4 he was President of the Welsh Liberal party. Then in 1974-5 he was **President of the UK Liberal Party**. Made a Life Peer in 1973. President of various organisations, such as The Foundation of Science and Technology, Mobile Radio Users Association. Vice President of a Parliamentary Scientific committee. Many Directorships in business.

Mr SAM GREEN, CBE, FIEE, FIProdE, FRSA. Died in 1996.
A Member since 1964, a Chairman and a Vice President. Chairman of IFIA. Awarded the **WIPO Gold Medal** in 1984. He is from Lancashire and worked in the textile industry. The inventor of the **box-motion in 4-colour automatic loom**.

He was awarded the CBE for work with the disabled and was MD of Remploy Ltd. In 1968 he arranged the meeting with Harald Romanus that started IFIA.

Mr LEONARD A. T. COTTERELL, MBE, MIPR, MAIE.

The General Secretary of the Institute from 1964 to 1977, covering phases 2 and 3. A highly organised individual with skills in many fields. His early career was spent in the Secretariat of the Royal Society of Arts. During WW2 he was a Staff Captain at Montgomery's HQ in 1944. Post-war he was in Personnel with Bowater Group. A Member of the British Association of Industrial Editors. He was also Vice Chairman of the Industrial Sports Club Secretaries Association.

Mr PATRICK A. RUSSELL-RAYNER, BSc, MInstP, FCIPA.

A Patent Agent, who spent almost a decade as the Secretary of the Institute. During the financially difficult times in the early 80s, he and ***Sam Green*** paid some of the bills out of their own pockets to keep the ship going. Now retired from almost all this sort of thing, but an advisor for the **Brian Wybrow** group as necessary. Since retirement, Pat has been bitten by the artistic bug and spends all his time away from the phone, painting.

Mr PERRY AKHURST.

A Member for many years and on the Council too in the early 90s. Although on the Council in the role of Business Manager, he did the Treasurer job a couple of years when no-one was available.

Mr MAX BEAUMONT.

He too is a long-serving Member and on the Council in the 80s. Treasurer during some difficult years, 1984-1990.

Mr ROBERT A. Von SYMES SCHUTZMANN, AOM: Comp.RAeS.

President of the Institute from 1991-1998, previously Chairman and a member from 1970. He left Austria in a hurry before WW2, and was in the Royal Navy as a Lt. Commander. From 1957 he was a presenter with the BBC, and became well-known on programmes such as Tomorrow's World and Horizon.

IPDC/IPI MERGER

In 1998 the **IPDC merged with the IPI**, to be known as the IPI, the object was to speak for all the inventors in the UK with a single voice and avoid duplication of effort. The IPDC was a national group of some 350 members, founded by ***Derek Myhill*** and ***Peter Knight*** in 1990. ***David Wardell*** came in as a Director in 1993. Later David and Peter instituted the magazine '***INVENTORS WORLD***'. All IPI members get a copy of the magazine as part of their subscription, and the IPI has several pages available in every copy for Institute articles and news. '***FUTURE AND THE INVENTOR***' continues, but less frequently.

A useful bit of publicity was provided at the merging ceremony when ***Mike Keenan*** demonstrated his invention for making materials non-flammable. The material was smeared on **Howard Stableford's arm** and then set alight with a blow-torch. Howard is still in one piece and is no doubt relieved and impressed.

In the latter part of the 90s, the Institute has supported Exhibitions and Innovation Competitions, such as the '**Great British Inventions and Innovations Fair' (GBIIF)** and the 'Tomorrow's World' Events.

After informing the Secretary about the significance of **September 13th** in the life-cycle of the IPI, Ruth said "that's the date of our next Council Meeting!". But of course everything went well enough.

At the **Lord Mayor of London's banquet**, held in November 1999 to honour '**The Best of British'**, about a dozen inventors were on the Guest List. Institute members prominent in this list included ***Lord Porter of Luddenham***, an Honorary Fellow and Nobel Prize winner in Chemistry (1967), ***Ron Hickman***, inventor of the **Workmate** and ***Trevor Baylis***, developer of the **Clockwork Radio**.

PROFILES (1998 -present)

Mr HOWARD STABLEFORD, BSc, MA ***President.***
He was a Geography graduate in 1977, but as a student indulged in some theatrical interests. Howard is from Lancashire and is married to an American, so he spends much of his time now in the USA. From 1982 he worked for BBC Radio Northampton. Then he performed (a well-chosen word) various roles in children's TV programmes. He was also a weather reporter, newsreader, etc, obviously an all-rounder and team-player. In 1985 he took a leading role in the new **BBC 'Tomorrow's World'** programme, and was simultaneously involved in the **Prince of Wales 'Young Inventor' Awards**. A particular success was winning an award at the Quebec International Science Film Festival for a film documentary about the **Chernobyl disaster**. Presently, he is still involved in TV but makes programmes on a one-off basis, and is diversifying. Howard has a media training company called '***Webzebra TV***'. One of his more exciting projects was making a programme called '***Animal Zone***' in the wilds in Ecuador. He runs the London Marathon for charity. In his role as President of the IPI he continues the deep interest in inventions and innovation that developed from his **'Tomorrow's World'** initiation.

Mr PAUL AMBRIDGE, BA, BSc, MSc. ***Chairman.***
Paul was a design engineer in his early years, principally with the automotive industry. In 1973 he joined the Design Council and progressed from product design evaluation to being the Southern Region Director in 1994. His success in advising on engineering designs prompted a great many patent applications and registration of new designs. He now has his own company PBA Associates, doing software development and programming. There are clients worldwide already. Paul has written a number of articles exposing the high costs of patents in the UK relative to rewards anticipated from royalties, compared to other countries. He has stockpiled a lot of valuable statistics on loss of trade to the UK by inventions going overseas.

Mrs RUTH MAGNUS. ***The Secretary.***

After a grammar school education, she took a bi-lingual secretarial course and language degree at the College Secretariat du Lycee Francais de Londres before going to live and work in Geneva. There followed 20 years of press and public relations work in the construction industry, usually with some linguistic aspects. In 1983, Ruth took up a position with the Common Law Institute of Intellectual Property as the Administrator's Assistant. She refers to this organisation as the **'other IPI'**. She was appointed Secretary of the Institute of Patentees and Inventors in August 1992. Ruth is likely to remain the only one who has been with more than one IPI. Ruth is also the Editor of ***'FUTURE AND THE INVENTOR'***.

Mr PETER KREBS, BSc. ***Hon. Treasurer.***

A Member for many years, Hon. Treasurer of the Institute at present. He is a physicist and has been in semi-conductor and microelectronics research at Mullard, Elliott, Plessey, Solartron and Ferranti. From the 70s he had much involvement with invention/innovation but never came out on the right end of successful items. However, his father's estate included some successful patents. So Peter continues to try and help other inventors in respect of his own good fortune from these sources. He offers a service to inventors by provision of **'Invention Marketing Letters'** and can include **marketing negotiation consultation**, which can be crucial to an inventor. There is much more about Peter's family in the related section later.

Mr GRAHAM JONES, CIPA. ***Deputy Chairman.***

A Fellow of the Institute. A European Patent & Trade Mark Attorney since 1971, with his own practice in London. He has a long history of service with the Institute, having joined in 1972 and has been on the Council a long time too. Much of his Council service has been as Vice-Chairman. He has always supported the Institute with his advertisements regularly in the magazine. One of his particular responsibilities has been as the IPI representative on the Standing Committee at the Patent Office. He is also an Honorary Advisor to the Prince's Trust for Youth Innovation. Among his clients have been the 'Inventor of the Year' in 1997, 1998 and 1999. Keep it up!

Mr ALAN WILCHER, IEng, MIAgrE, MIRTE.

A Fellow of the Institute. ***Executive Management/Council Member***.

On the Board of Management of the IPI Executive Council from 1988. A project engineer whose inventive background and successes are well detailed in the section dealing with inventors later. Alan has 40 years experience in management of invention, innovation and product development, mainly in Agricultural, Civil Engineering and Transport industries. His know-how extends through production to marketing and negotiation. Within **IFIA**, Alan is the Youth Advisor representing the IPI. Alan was also the Principal Founder and driving force behind the **Association of Innovation Management (AIM)**, which in 1996 was subsumed into the IPI. His extensive contacts at higher levels, in a wide variety of industries, means his advice is invaluable to an inventor at the stage of trying to commercialise the product.

OFFICERS OF THE INSTITUTE OF PATENTEES & INVENTORS

Paul Ambridge, Chairman.

Howard Stableford, President.

THE AUTHOR

Derek Myhill, Historian.

Mrs Ruth Magnus, Secretary

Graham Jones
Deputy Chairman

Peter Krebs, Treasurer

Alan Wilcher

The IPI Council Members after the merger
of the IPI and the IPDC

Jonathan Copus

David Wardell

Peter Knight

Mike Watson

Kane Kramer

THE IPI OFFICE
(Ruth at work)

The imposing facade of No.10 Victoria St.

The Lecture Hall (200 seat)

Library and reading rooms

'INVENTOR'S CORNER' Bar
(Ladies not admitted)

A committee or meeting room

Mr S.E. Denning
Treasurer

Capt. G. Drury Coleman
Secretary

Lord Askwith
Pesident

Professor P.M. Baker

Sir William Bull
('Whats Wanted')

Professor Low
(and assistant)

THE JUDGING PANEL

EXHIBITION (1930)

The IPI stand at LINPEX.

The Director of the CBI,
Mr A.G. Norman CBE, with an exhibitor.

A view of the LINPEX exhibition.

Sir John Maitland (President) with Rt. Hon. Tony Benn at the dinner.

Right Hon. Tony Benn,
presenting awards at the dinner.

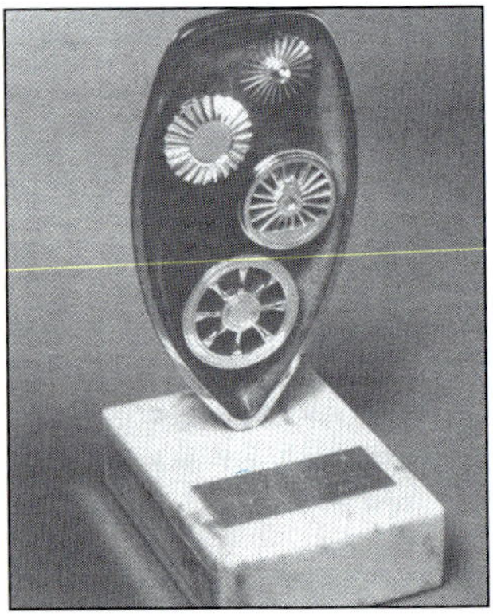

The Financial Times Trophy.

Sir John Maitland
President (1974).

Mr Arthur W. Richardson
Chairman (Gold Medal Fund).

Lord Lloyd opening the 60th Anniversary meeting at the Royal Society 1979.

David Mitchell MP with Lord Lloyd at the 60th Anniversary exhibition.

IPI HOUSE OF LORDS RECEPTION
(1974)

Sir John Maitland (President) welcoming Lord Sherfield

Lord Inglewood (IPI Vice President) welcoming The Right Hon. Geoffrey Howe, the guest of honour.

1925 Exhibition.

№ 170887

Certificate of Incorporation

I Hereby Certify, That the

Institute of Patentees (Incorporated) (the word "Limited" omitted by Licence of the Board of Trade)

is this day Incorporated under the Companies Acts, 1908 to 1917, and that the Company is Limited.

Given under my hand at London, this Eighteenth day of October One Thousand Nine Hundred and Twenty.

Fees and Deed Stamps £ 22. 10/-

Stamp Duty on Capital £ —

Registrar of Joint Stock Companies.

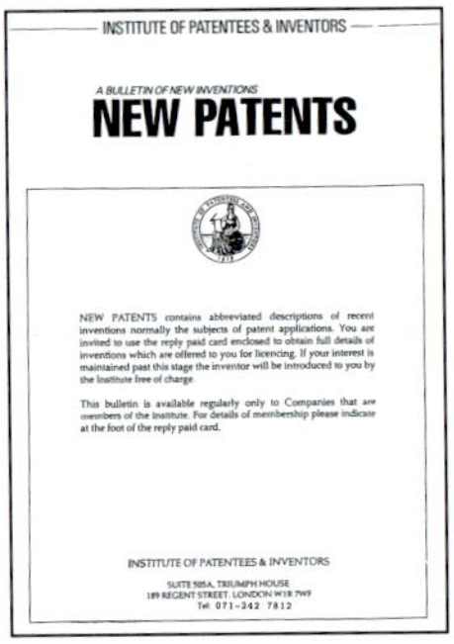
INSTITUTE OF PATENTEES & INVENTORS

A BULLETIN OF NEW INVENTIONS

NEW PATENTS

NEW PATENTS contains abbreviated descriptions of recent inventions normally the subjects of patent applications. You are invited to use the reply paid card enclosed to obtain full details of inventions which are offered to you for licencing. If your interest is maintained past this stage the inventor will be introduced to you by the Institute free of charge.

This bulletin is available regularly only to Companies that are members of the Institute. For details of membership please indicate at the foot of the reply paid card.

INSTITUTE OF PATENTEES & INVENTORS

SUITE 505A, TRIUMPH HOUSE
189 REGENT STREET, LONDON W1R 7WF
Tel: 071-242 7812

Herbert N. Casson, Editor.

THE LIGHTER SIDE

It's a device for transferring software to make it patentable

"I do feel so completely ridiculous, Mr Stephenson"

The IPI stand at
Tomorrow's World Live 1999

Bob Symes' retirement award presentation
(1998); from left:
Paul Ambridge,Howard Stableford, Bob Symes

A view of Staple Inn
Buildings - IPI office
location in the 80's.

MEDALS & CERTIFICATES

INSTITUTE OF PATENTEES
AND INVENTORS

Certificate Of Membership

This is to certify that

has been admitted as a

Member

of the Institute

Given under the Common Seal of the
Institute of Patentees and Inventors.

Peter Blanchard
with 'Blind Spot Mirror' and 'Volcano'.

Jonothan Copus
receiving his 'SMART' award.

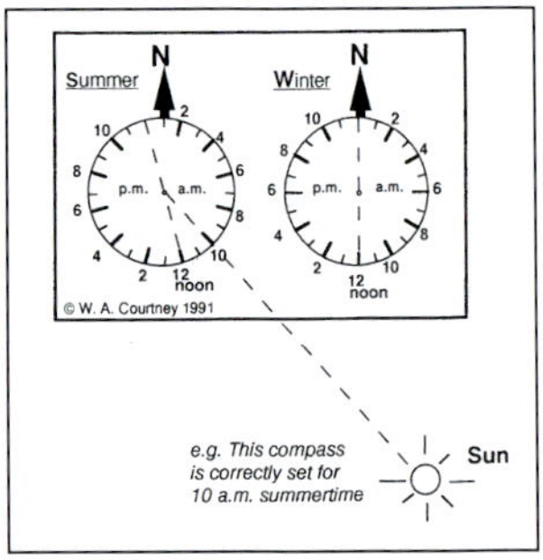

William Courtney's
Solar Compass.

The THAMES BARRIER,
designed by Charles Draper.

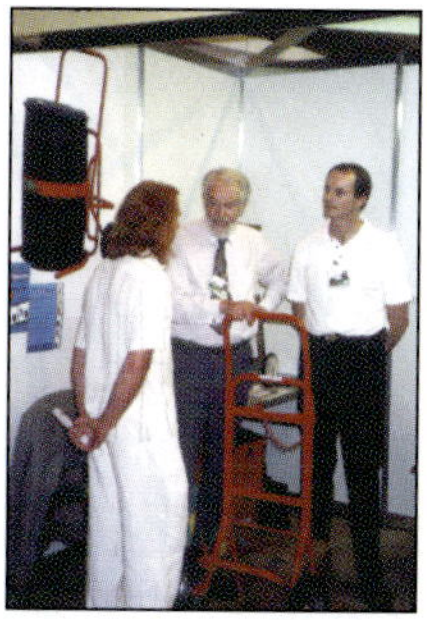

Henry Crawford,
exhibiting his upright trolley.

Ron Hickman, with
the 'WORKMATE'™.

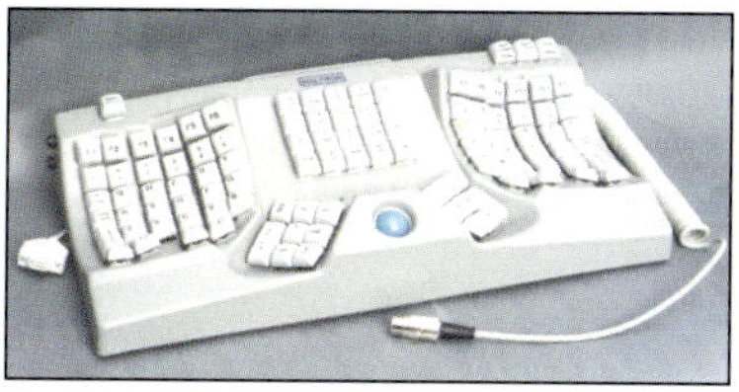

Stephen Hobday's
MALTRON keyboard.

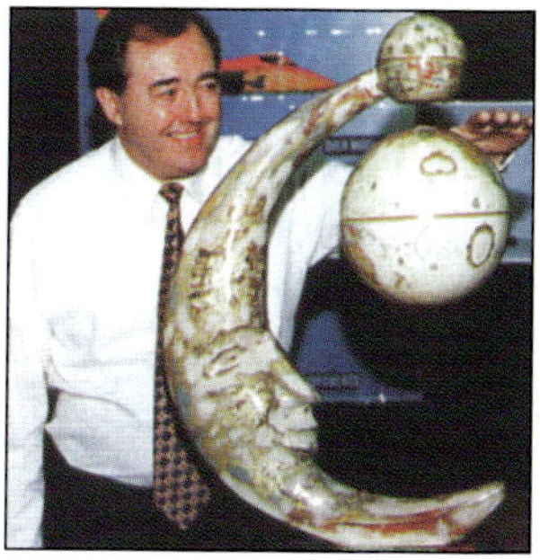

Barry Jones, and the
magnetic suspension device.

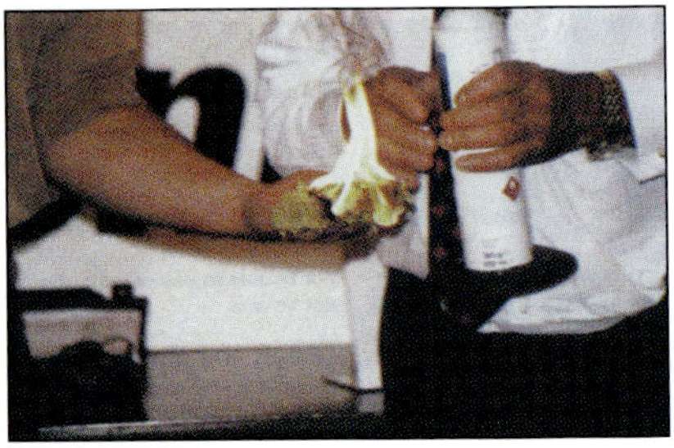

Mike Keenan's
'non-flammable' grease
(on Howard Stableford's hand)

Photo© Matthew Thorne

Barry Langrish's
'Wine-Stems' on display.

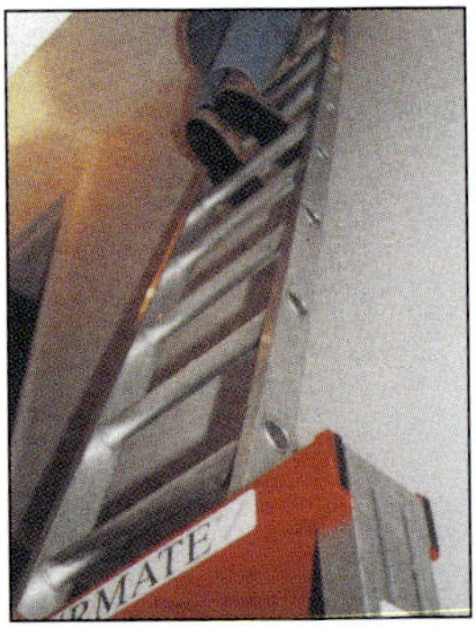

'STAiRMATE' invented by
Stan Lewis and John Standham.

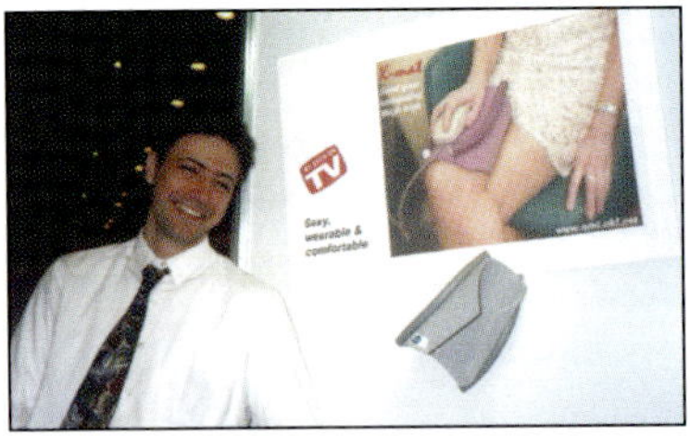

Nicholas Mark
Exhibiting his 'K-mat'.

Joe Michael's
Shape-changing robots.

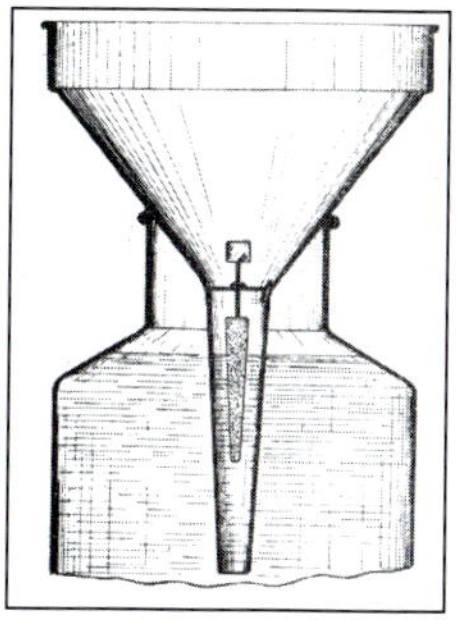

Miss Miller's
'Jack in the funnel'.

Derek Myhill's,
plant 'MINICULTURE' system.

Tim Snape, with the
Portland underwater horn.

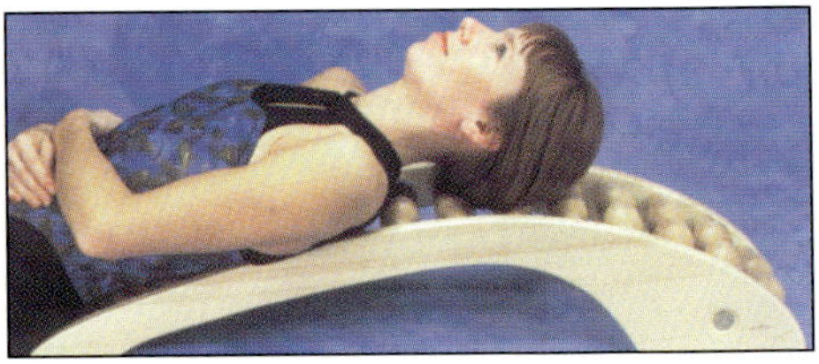

Neil Summers'
'BACKSTRECHER' at work

Andrew Warren, demonstrating
'SLIDE-SAFE' on a sliding door.

Iris Whyte, with the
'TALKING POTTY'.

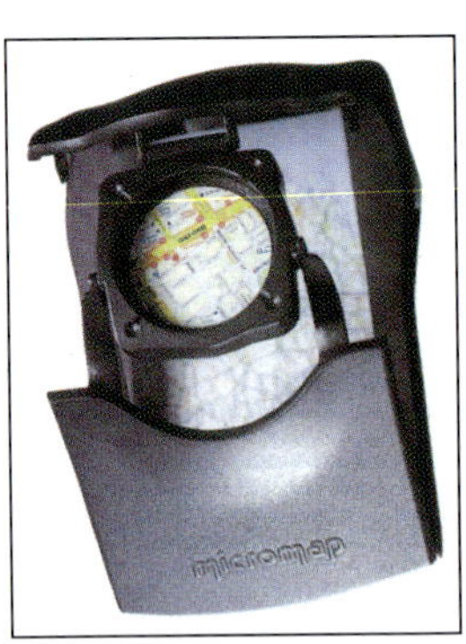

Jeffrey Woolf's
'MICROMAP' System.

REVEREND JONATHAN COPUS. **Council Member.**

The founder of Dentron Ltd, 10 years ago. This is a Medical Electronics company to produce and market his inventions. He has been successful in obtaining grants for research and innovation development, including 3 DTI 'SMART' awards. His **'BIOGUN'** invention is well described in the inventors section later. He is highly versatile and now operates a TV and Radio Services company called 'IMMEDIA', to train young people in maximising benefits of usage of the media. Jonathan uses the term **'grantsmanship'**, whatever it is - he has it! Hopefully he can pass on the secret to other inventors via his activities as a Council member.

Mr PETER KNIGHT, BSc, CEng, MBCS. **Council Member.**

A Fellow of the IPI from 1998, and also co-opted onto the Council following the IPDC/IPI merger. From 1999, a regular Council Member. Peter was a Founder Director of the IPDC, and then a Director of Inventors World. His inventorship details are listed in the relevant section later. At the Inventors World, Peter's responsibilities are mainly as Research Director, which suits his IT education and experience. He is a bit of an 'ideas man' anyway.

Mr KANE KRAMER. **Council Member.**

He had an early training for the entertainment world, but entered the family business which was in footwear and clothing. He has memories of taking things apart from the earliest age, but to see how they worked. From 1979, he went back into entertainment business where he developed a product which he called **'IXI'**, the product got nowhere, but the name stuck. Later, he and his wife went into retail furniture in Hertfordshire, expanding into exclusive kitchen design work. One project required matching colour/details for appliances. The end-result was the invention of **'METALCOAT'**, a water-soluble metallic paint. This has been a big success story, detailed in the inventors section. Guess what the brand name is? On the Council since 1998, contributor to the Radio Cambridge 'Inventors Phone-in'.

Mr DAVID WARDELL. **Council Member.**

Co-opted onto the Council in 1998 following the merger of the IPDC into the IPI. Now a full Council Member from 1999. David is the Editor in Chief of Inventors World. He has a diverse background in Catering and Hotel Management, Manufacturing Industry, Sales, Marketing, Insurance/Finance. In 1993 he joined the IPDC as a Director, in 1998 the IPDC merged into the IPI, in order to create a single national voice for all UK inventors. A Fellow of the Institute. His services to the invention industry are detailed in the inventors section later.

Mr MIKE WATSON. **Council Member.**

A Council Member for some years now. He was educated and trained as an electromechanical engineer, early work being in industrial design and development. Mike is from Surrey, and has mainly been self-employed. He describes himself as an **'inventrepreneur'**. A prolific inventor with diverse work interests. His most well-known invention, the **'Watsonlinc'**, TM, a unique telephone recording device is detailed in the inventors section later. Mike has strong views about how ideas generators (inventors) need the wealth incentive just like the rest of society.

This ends part 4 of the IPI story, but it is really the beginning.

********** Quaint but curious / the lighter side **********

Obviously from time to time the Institute has had to tactfully explain that an invention submitted was not needed by anyone, cannot work or even be manufactured. Over the centuries, some very humorous applications have accumulated at Patent Offices. Here are a few for your entertainment.

In 1718, Mr Puckle patented a **'Portable Gun'**, which would '*Fire square bullets at Turks and round bullets at Christians*'. A description on drawings said 'Defending King George your country and lawes, is defending yourself and Protestant cause'.

In 1794, Garratt describes **'Queen's Cocoa'**, which consists of:
112 lb. of dried coconut, 113th part of orange peel, 114th part of lemon peel, 115th part of cassia, 116th part of Peruvian bark.

An 1866 patent for **artificial tea** by Davis. Bichromate of potash is decomposed with sulphuric acid, then chromic acid and ammonia added. When the solution is evaporated and crystallised the product resembles tea.

In 1869, Michel designed an **aerial machine** comprising a flock of birds harnessed to a triangular frame, from which is suspended a car. Reins control the birds and rudders control the car. *Patent was refused.*

In 1901, Chopard & Chopard, had a method of capturing **pure air** from a healthy locality and transport and release in an unhealthy locality. Said air as distributed was warm or cold, containing antiseptics, perfumes, etc.

In 1903, a German designed **phonographic plates made of chocolate**, after playing the record you ate it.

In 1853, an **improved bedstead** by Eliezer, 'which also may be used as a **vehicle**'.

Also in 1853, John O'Leary designed a **chest** for the use of emigrants, 'which could alternate in use as a **wheelbarrow** and b**edstead** too'.

In about 1930, one of the 'What's Wanted' was a
'**gas lighter than Hydrogen, for some aeronautical purpose**'.

Even in 1965, in the USA, an inventor of some specialised handles for a wheelbarrow conducted a patent search and discovered there was no patent for a wheelbarrow. So he filed for the **wheelbarrow** itself and was sitting back waiting for royalties!

WOMEN MEMBERS OF THE INSTITUTE

Although tending to keep a low profile, there have been women members since the 1920s. ***Lady Askwith*** once addressed the Annual Dinner. ***Dame Caroline Haskett, DBE***, the Director of the Electrical Association for Women, was a Council member in early days. She had been a judge at the first annual exhibition in 1924. Quite a few women have been Council Members (if records were intact no doubt more would be evident). ***Miss M. M. Turner*** was on the Council around 1970-2, and in 1979 ***Miss Mary Vitoria*** and ***Miss M. Bennett-Robinson*** were on the Council together. Miss Turner had Fellowship status, as does ***Miss Georgina Richards*** currently (a Patent Solicitor with the Southampton University Intellectual Property Unit). Georgina is also a member of the National Innovation Centre Steering Committee. Miss Bennett-Robinson was an inventor and exporter of new products.

There is now a competition for **'The Woman Inventor of the Year'** in the UK, but even in the Institute's earliest exhibitions there was always a section for women. An outstanding inventor in 1929 was ***Miss D. B. Miller*** (Isle of Wight) who devised a filter funnel with a removable flotation device inside, which made filling bottles easy without overspills. Why isn't it still available? An enthusiastic IP Member, she was taken aback at the great deal of help she got from the Institute after winning a medal at the exhibition. In 1998, ***Bola Olabisi*** organised the first UK 'Woman Inventor of the Year' competition, won by ***Mrs Iris Whyte*** a Member of the IPI.

Several women have of course performed important management roles for the Institute. In the early 1960s ***Mrs Hazeldine*** (the longest-serving employee then, performed the role of General Secretary temporarily for three or four years. When she retired in 1974, having been there 45 years, she was the longest-serving employee. ***Mrs Enid Cordeiro da Silva***, was the Assistant Secretary in the early 70s a title used then for Editor of the magazine. Mrs da Silva had previous experience with a publishing company and so was quite valuable in that role. An indication of her significance in the Institute was that she was a guest at the House of Lords reception in 1974. As outlined above, ***Mrs Ruth Magnu***s performs two roles at present, that of Administrator of the IPI offices and also that of Editor of the magazine. Ruth performs both functions admirably, and she will be pleased to learn that other women have played important roles in the past.

Some women members in the 30s did make good incomes from their products, but kept out of the limelight. Inventions successful commercially were:-

a special spoon, pram-tray, cuff-purse, travelling wardrobe, vegetable stand, and a shoe-rack. One woman evidently filed 20 patents in a single year. The big one seems to have been the invention of the cardboard matches in soft folders, some woman did get rich on that one, but who was she?

More evidence of the existence of women members early on is that the annual competition for a poster design for the international exhibition was often won by a woman member. One individual, ***Mrs D. Penna*** (Surrey) was conspicuous on picking up the prize money on several occasions.

THE FAMILY AND THE INSTITUTE

When checking back with older members, trying to fill in the gaps, it became plain that young people take up the parents' interests in inventing. Here are a few of the family involvements beginning to surface.

John Low, son of ***Professor Low***, was a member for 50 years and attended meetings regularly up to his death in 1981 (25 years after father died). But ***Professor Low's father*** started it all, since he was a member of the Institute in 1920, attending meetings in Gt. Russell Square even before his son the Professor.

Mr H. F. Webb, LRIBA, was a member from 1922 and a long-serving Council Member (Vice President) until his death in 1952. His son ***Mr E. H. C. Webb*** had become a Council member too, from the year prior to his father's death. Mr Webb, Sr, was in 1930 leading the search for premises for an 'Inventors Hall', a site had been identified (on a road-island) in Central London, and a donation of £5,000 was available towards it. Much larger than needed but extra space would have brought in extra revenue. It is not known why it did not proceed. The Institute is trying to trace members of that family to get more details.

In 1952, ***Mr A. J. Wilkinson***, the inventor of **'LINATEX'** (a new type of process rubber), decided to reside in South Africa. He had been a Fellow from 1930, and on the Council many years. At the time of his departure, Wilkinson, Jr was ready to join the Council.

Ron Marchant, who is currently Assistant Comptroller of The Patent Office, and well known at the Institute, is proud of his father's medals won in IPI exhibitions in the 60s for Jam-making processes and wheelchair improvement devices.

Arnold Tickner, FCIPA, a former Council Member of the Institute, was also a nephew of a previous Assistant Patent Comptroller.

The present Treasurer Peter Krebs has interesting details of his father's Institute involvement. His father ***Wolfgang*** was a Jew, born in Germany during WW1. By 1935, he had a science degree and was working in the Patents Dept of AEG Company. Friends advised him to get out of the country, but although knowledgeable on electrical rotating devices he was unable to publish in scientific journals. So he wrote technical articles in the company's magazine (which were seen overseas). Eventually he got job offers elsewhere. By 1936, he was in England (furthest from Germany), employment was difficult, being German-born, but he was safe. Next came WW2 and he was interned on the Isle of Man (a holiday?). In the confusion he forgot to put the bank account in his wife's name, the account was frozen and she had a hard time. By 1946, he was working in London, had joined the Institute then and commercialised several patents. Wolfgang became a Consultant and was known for designing a large special winch for the Admiralty. What is not commonly known is that Peter's uncle was ***Sir Hans Adolf Krebs***, the *Nobel-Prize* winning biochemist, after whom the **Krebs Citric Acid Cycle** is named. Hans had taken a similar route from Germany (much earlier than Wolfgang) and became Professor of Biochemistry at Oxford. His **Nobel Prize for Medicine** (1953) was for discovery of Acetyl-Coenzyme A, and the tricarboxylic acid cycle by which carbohydrates are metabolised. What a family!

Mr C. O. Forestier-Walker, a prominent Council Member for many years, was also from a family with several generations of inventors.

Alan Lovegreen, a Council Member in the late 70s, simply states that he was from a family of inventors.

Eric Hardman Taylor was a Member from 1951 and later a Fellow. After his death the family business was then run by his son ***John*** (who was also a Fellow).

The last relationship to be discovered is fascinating because the Institute is now able to provide additional details to someone about his ancestor. Knowing that the name Grey-Wilson is quite rare, enquiries were made of ***Mr Piers Grey-Wilson*** as to whether he was related to ***Sir William Grey-Wilson*** (one of the 7 subscribers to the Memoranda and Articles of the Institute when incorporated in 1920). Piers said yes, Sir William was his great-grandfather, and he knew that he had been involved in the Institute because a family historian discovered that the Institute had a Grey-Wilson Medal (for the best patent of the year). However, I can tell him that Sir William was a key member, since he had been the President and was in his second year of Chairmanship at the time of death.

No doubt now the subject has been mentioned the Institute will be hearing about other 'inventor-families'.

OLD, BUT DEFINITELY NOT 'OUT'.

A review of the obituaries suggests that inventor-members are typically quite old at the time of death. Several have been in their nineties and a couple over the century. Usually they are still working on some invention at the time. ***James Robb***, a Scotsman, died in 1996 at the age of 98 having been a Member for 60 years.

However, no guarantees are offered to potential members! Several present Members come immediately to mind.

Derek Richard Stuckey, in Berkshire, joined in 1943.

Henry B. Crawford, actually first joined the IPI when he was 74. And he takes good advantage of the reduced membership fees for the elderly. In the 1999 'Tomorrow's World Live' show he was exhibiting his invention, a 2-wheel trolley which can carry a heavy load upright. He was 83 at this stage. His organisation is Level-Lift Trolleys Ltd, and he has at different times referred to himself as 'young

70s plus' and 'somewhat advanced age'. In fact he had been seriously concerned that he would not be allowed to exhibit at his age.

Another 'old youngster' is ***Louis Anthony Ralph Ross, CEng, MICE***. A civil engineer in Woking, he has loads of inventions including various clamps for use with portable workbenches, oil-spillage retention barriers which are easily transportable to spills at sea, gas-leak detectors, etc.

Stephen Hobday, inventor of the Maltron Keyboards is elderly, as is ***Max Beaumont***, both are active Members. Max is currently working on a new chimney system, but says it is a modification rather than an invention.

MEMBERSHIP STRUCTURE

MEMBERS:
The main body of the subscribers. Inventors or innovators, entrepreneurs, Patent Agents/Solicitors.

ASSOCIATES:
This class of membership was for individuals who were not inventors but wished to be aware of developments for personal reasons. In the past the objective seems to have been to accept them at a low subscription rate and then in time convert them to ordinary full members. They did not use the general services and probably were only wanting the journal, so cost little to accommodate anyway. From 1999, entry to the Institute is through the Associate level which can be upgraded on request.

MAKER MEMBERS/COMPANY MEMBERS:
The old-fashioned term for manufacturers whose main interest was in knowing about licenses available for new inventions/products. In 1975 the terminology was changed to COMPANY MEMBERS. They pay a higher fee to belong and be among the first to hear of new developments.

STUDENTS/COMPANIONS:
There has always been some facility for students to get the journal at a low fee and have some services but not a vote. The Companions term has been used in several contexts but currently is a class of membership for youngsters, below university age.

FELLOWSHIPS:
There are currently 20 plus Fellows, which includes 3 with honorary status.
The rules have varied over the years, but one principle has not changed. This being that unless a Fellowship is Honorary, it cannot be awarded at the time of applying for Membership. Generally a number of years membership is normal before an offer of Fellowship might be made to the individual. In earlier days a successful inventor could apply and submit a thesis concerning his invention. A panel of judges selected by the Institute would then evaluate not only his product design merits but also the applicant's understanding of intellectual property principles, and also commercial development and the technical or scientific principles relevant to the type of product. The judges would then present their recommendations to the Council. A majority vote was still required from the

Council for Fellowship award. This mechanism is no longer relevant. Honorary Fellowships now represent the equivalent of the previous Vice Presidencies. In general Fellowships are now awarded in recognition of successful completion and commercialisation of a significant new product concept, or as recognition of a long service to inventors in general or to the Institute specifically. It was at one time considered equivalent to a degree in **'Inventorship'**, but such ideas are no longer popular, and the modern intent is purely to acknowledge the **lifetime contribution** of an individual to the world of invention/innovation.

VICE PRESIDENTS:

These in the past were both general and honorary. The honorary was for important individuals in the judiciary, parliamentary and sometimes industrial sector who had demonstrated considerable interest and skill in matters relating to intellectual or industrial property. It was expected that their efforts on behalf of the Institute would benefit inventors in the UK by influencing political reality on the subject. Most Vice Presidents were long serving and productive for the Institute's objectives. Many continued until forced retirement or early death. The mechanism is not now in use.

LIFE MEMBERS/LIFE FELLOWS:

A number of Members take advantage of preferential rates and take out a lifetime subscription. There are more than 30 such, including:
Robert Redding, FIEE, MBCS, FInstF, (Fellow), ***Professor M. Thring, Mrs Maureen Bennett-Oates*** in Vancouver (B.C.), ***Ronald Woolf,*** in Durban (S.A.), and ***Douglas W. Nield, FID***.

LOYALTY/ALLEGIANCE

The institute has made enormous efforts to cater for each member's specific needs, and there have been no splinter groups or divisions of any kind. In the author's experience this is quite remarkable for such a society. However, the Institute has always demanded a high level of loyalty and allegiance from its members in return. Another aspect of this is the high degree of commitment from everyone at the management end. Most executives are parted by death rather than retirement. ***Mrs Hazeldine*** is a particularly good example. It would be a fairly good bet that she was married to a prominent barrister, and was not working at the office from necessity. She 'retired' in 1959, after 30 years service as the senior secretary/typist, only to be recalled as the temporary General Secretary within a year and then became the Company Secretary for several years. It was actually 1974 before she got away, after 45 years service.

In this context, the existence of so many families in which successive generations are active in the Institute throughout their adult life speaks louder than words.

THE INSTITUTE AND THE FUTURE

Approximately every 20 years there have been articles in the Institute Journal querying where the future lay. Undoubtedly such a period is being encountered right now, coincidental to the Millennium.

There was a suggestion in 1992 by the then Chairman, ***Mr Bob Symes***, that a change of name might be in order, '**The British Institute of Patentees & Inventors**' was proposed and generally found acceptable. But current advice was that a fair amount of expense would be involved and the time to do it was not just then. There may be a good case for a return to the traditional title for the magazine, it is unlikely a more descriptive name could be found for the house journal.

This has not been a tale of 2 cities (except for 1937). But it certainly is a tale of 2 wars, 3 industrial revolutions, 1 street, many magazine styles/names, lots of logo changes, several organisations, a number of key management, about 200 exhibitions for inventions, and 30,000-40,000 members with their individual projects and problems. As Paul Ambridge points out in his foreword, the number of people helped by the Institute during the 80 years is several hundred thousand.

One might consider that 4 different ages of industrial revolution and hence invention have been experienced throughout these 80 years. The first was clearly of a mechanical nature, the second very electro-mechanical, the third has been and is electronic. But unlike previous phases the electronic age is concurrent and overlapping with a fourth 'virtual' phase (and facilitates it). ***David Nicholas***, of Hampshire Business Link, has devised **'The Virtual Company'**, and what with virtual reality, virtual actuality, and virtual shopping over the Internet, where will it end, maybe with a 'virtual' patent.

The **European Patent Convention** and **The International Patent Treaty** are both now well understood by experienced inventors, there is less demand for information relating to these patent systems. In addition, the ready availability of all kinds of information via the **Internet** (including **Patent Search** mechanisms) means many of the more capable inventors go it alone in the early stages. There may be a tendency not to join the IPI until they have patent applications in place and probably have a prototype. Then when it becomes a matter of making contacts they suddenly realise the need to belong to such an organisation. Members mainly stay in the Institute once they have joined but may be joining at a later stage than in years past.

Without any doubt, Institutes and Societies throughout the developed World all face similar problems. What services should they provide for members in the coming era? What will the members still want, or not need any more?

Recent UK statistics suggest there are still some 50,000 viable ideas surfacing in the UK every year. Most of these inventors spend a lot of money on prototypes, patents, advice, research and development, the average spend is believed to be about £5,000 per individual. This would be a total of £25 million for the 5,000 who apply for a patent, which is many times greater than the total funds available from the government in any year to support or encourage invention and innovation. Why do I find this surprising? Who knows what the other 45,000 inventors spend before giving up?

One aspect of there being so many ideas floating about is that there are ever more ways for a lone inventor to get information together, but they phone the IPI anyway and consume a lot of time out of the administrator's schedule. Then the caller goes off and tries the Internet, calls Inventors World or Business Links, etc. The Institute being well known and established means that government departments refer queries to the IPI. It is nice to be well known but this 'Public Service' aspect consumes valuable funds. Whenever the DTI gives a grant or form of funding in recognition of this, the amount is inadequate for the purpose. Much of the time there has been no help. In 1996, ***The Hon. David Shaw, MP for Dover***, addressed Parliament on the subject of support for inventors. He drew attention to the enormous number of enquiries from the general public, and the fact that the government had then granted £10,000 towards the cost of this public service aspect of the Institute's activities. While pleased that the Institute was getting some funding he considered the DTI could easily afford to do better than that.

CONCLUDING REMARKS

The main point is that there are an awful lot of people out there trying to progress their own **'Bit of an Idea'**. If inventors are all a bit strange, it's OK because there's a hell of a lot of us! So since 1919 what is new. *NOTHING!* Each successive government has had to be re-educated by the Institute to the needs of inventors (this is regardless of the party in power), it is a never-ending battle. What is also not new is that the Institute is the only representation for the inventor, otherwise their voices would not be heard at all. The need for the Institute has and always will be there.

The IPI has thus filled a void left by all other professional official bodies in the UK. It has been the only source of unprejudiced information and liberal guidance on intellectual property and industrial property rights, and a genuine commercialising service for the promotion and placing of inventions.

With your continued support, long may it be so!

OFFICERS OF THE INSTITUTE

PRESIDENT:

1919. Sir Joseph Lawrence, Bart.
1920-1922. Sir Harry Seymour Foster, JP, DL.
1923. Mr Morgan Morgan, KC.
1924. Sir William Grey-Wilson, KCMG, CMG, KBE.
1925-1942. Lord Askwith, KCB, DCL.
1942-1956. Professor Archibald Montgomery Low, DSc, ACGI, FFSc, FICA, MIME, MIAE, FCS, Hon. FInstPI.
1957-1961. No President.
1962-1966. Sir Stanley Rawson.
1966-1975. Commander Sir John Maitland, RN (Ret), JP, FRSA.
1975-1990. Lord Lloyd of Kilgerran, CBE, QC, JP, MA, BSc.
1991-1998 Mr Robert A. von Symes-Schutzmann, Comp. R.AeS.
1998 - present. Mr Howard Stableford, MA, BSc.

CHAIRMAN:

1925-1926. Sir William Grey-Wilson, KCMG, CMG, KBE.
1956-1974. Mr Arthur Whittaker Richardson, OBE, FInstPet, FCS, FRSA, Hon. FInstPI.
1975-1981. Mr Sam Green, CBE, FIEE, FIProdE, FRSA.
1982-1986. Dr L. L. Ware, OBE, PhD, MA, BSc, ARCS, FRIC.
1986-1991. Mr Robert A. von Symes-Schutzmann, Comp. R.AeS.
1991 - present. Mr Paul Ambridge, BA, BSc, MSc.

VICE PRESIDENTS:

1921-1930. Mr Morgan Morgan, KC.
1925-1960. Lord Brabazon of Tara, MC, MP, GBE. (Hon. VP).
1926-1942. Professor Archibald Montgomery Low, DSc, ACGI, FFSc, FICA, MIME, MIAE, FCS, Hon.FInstPI.
1926-1939. Mr W. J. Dow, FCIPA.
1929-1930. Sir Richard Ashmole Cooper, Bart, MP.
1929-1933. Mr George Herbert Skinner.
1934-1955. Sir John McDonald, JP.
1936-1939. Capt. The Hon. V. A. Cochrane-Baillie, MC.
1936-1953. Mr H. F. Webb, LRIBA.
1936-1955. Mr C. A. Shierwater, MBE.
1948-1957. Capt. Sir Arrol Moir, Bart, BA, MInstCE.
1951-1955. Mr S. E. Denning, FCIS, FSS.
1955-1956. Mr B. Gardner, OBE.
1962-1963. Major C. W. de Roemer, JP, Comp.IEE.
1962-1970. Mr C. D. Goodhart.
1964-1970. Mr R. Welton Cook, MIED, FRSA, Hon. FInstPI.
1964-1975. Mr L. Gordon Davies, Hon. FInstPI.
1965-1984. Mr Arthur Whittaker Richardson, OBE, FInstPet, FCS, FRSA, Hon. FInstPI. (Hon VP).
1966-1973. Sir Harry Legge-Bourke, KBE, DL, MP. (Hon.VP).
1966-1975. Lord William Morgan Inglewood, TD, DL, MA, ARICS, MP. (Hon. VP).
1968-1981. Mr Sam Green, CBE, CEng, FIEE, FIProdE, FRSA.
1972-1981. Mr O. W. Lipton, BCom, MInstM.
1973-1981. Rt. Hon. Sir Richard Napier Luce, BA, DL, MP. (Hon. VP).
1974. Lord Lloyd of Kilgerran, CBE, QC, JP, MA, BSc.(Hon VP).
1975-1977. Mr Bill de May, FRSA, Hon. FInstPI.
1977-1991. Lord Edmund O. Ironside, FCGI. (Hon.VP).
1978-1979. Mr Alan T. Lovegreen.
1978-1982. Mr Owen Nutt, FRSA, FSCT, Hon. FInstPI.
1981-1991. Dr L. L. Ware, OBE, PhD, MA, BSc, ARCS, FRIC.
1981-1991. Mr O. J. Orchard, CEng, MIEE, CPA.

SECRETARY:

1921-1925. Captain G. Drury Coleman.
1926. Mr Morgan Morgan, KC.
1927-1940. Captain G. Drury Coleman.
1940-1948. Capt. Sir Arrol Moir, Bart, MA, MInstCE.
1948-1954. Major Frederick Charles Beauchamp.
1954-1960. Major R. O. Sutherland.
1960. Lt. Col. Ambrose John Crawley (acting).
1961-1963. Mrs M. D. Hazeldine, Hon. MInstPI.
1963. Lt. Col. Ambrose John Crawley (acting).
1964-1977. Mr Leonard A. T. Cotterell, MBE, MIPR, MAIE.
1978-1979. Mr Alan T. Lovegreen.
1979-1987. Mr Patrick A. Russell-Rayner, BSc, MInstP, FCIPA.
1987-1992. Mr John R. Kay, BSc.
1992 - present. Mrs Ruth Magnus.

EDITOR OF THE JOURNAL:

1930-1937. Mr J. V. Morton. (Editor).
1938-1939. Mr Herbert Casson. (Editor/Publisher).
1971-1975. Mrs Enid Cordeiro da Silva, BA. (Assistant Secretary).
1975-1978. Mr Grant Nightingale. (Assistant Secretary).
1979-1987. Mr Edward Gear. (Assistant Secretary).
1987-1992. Mr John R. Kay, BSc. (Secretary).
1992 - present. Mrs Ruth Magnus. (Secretary).

TREASURER:

1920-1924. Mr Joseph Ponsonby Horatio Bewsher.
1925-1955. Mr S. E. Denning, FCIS, FSS.
1956-1968. Mr E. H. C. Webb, FIAC.
1969-1977. Mr Thomas Arthur Negus, BA, ACA.
1978. Dr A. G. Thompson, BS, MB.
1978-1979. Dr L. L. Ware, OBE, PhD, BSc, MA, ARCS, FRIC.
1980-1981. Mr Owen Nutt, FRSA, FSCT, Hon. FInstPI.
1982-1983. Mr Sam Green, CBE, FIEE, FIProdE, FRSA.
1984-1990. Mr Max Beaumont.
1991. Mr Perry Akhurst. (Business Manager).
1992. Mr Andrew Chipling.
1993-1994. Mr Perry Akhurst.
1995 - present. Mr Peter Krebs, BSc.

THE INSTITUTE OF PATENTEES AND INVENTORS 1999-2000

President	Howard Stableford
Chairman	Paul Ambridge
Secretary	Ruth Magnus
Treasurer	Peter Krebs

Executive Council

Jonathan Copus	Graham Jones
Peter Knight	Kane Kramer
David Wardell	Mike Watson
Alan Wilcher	Jeffrey Woolf

SOME OUTSTANDING MEMBERS

Famous Members, Fellows, Council Members and Vice Presidents.

Mr Derek C. Adams.
A Fellow of the IPI, a Member since 1969. From Lymington, in Hampshire, he has formed an active local group of inventors that meet there monthly. The group is called MIRAGE, which represents Model Makers, Inventors, Researchers, and General Engineers.

Mr Paul B. Ambridge, BA, BSc, MSc.
Present Chairman of the Institute. Initially an engineering designer with TRW Cam gears, then in automotive design research. In 1973 he joined the **Design Council**, evaluating products submitted in design competitions. Later he became a Design Advisor helping companies to solve design problems in new product development. Paul's suggestions to them resulted in many successful designs and led to additional patent applications. In 1994, he had progressed to Southern Regional Director for the Design Council. Since then he has developed his own enterprises, trading as PBA Associates, a consultancy dealing with software development and programming work. The business has flourished and now has a number of government departments as clients. From 1988-1993, he organised the **Toshiba Year of Invention** competitions, some 20,000 entrants were considered. Paul has served as a judge and chairman of the judging panel for various international invention exhibitions. He has also written leading articles on a number of topics relating to the invention and innovation processes.

Mr Branko Babic, BSc.
A chemist, who has had some intriguing research projects. Branko is a prolific inventor, who has unfortunately not often ended up with the financial rewards associated with widespread use of one's inventions. Soon after higher education, he invented the **'vice' adjustable spanner** which was not taken up but was very effective. In 1985 he devised a **'mast to boom fitting'** for boardsailing. It is a method of securing the sail to the mast, and is sold worldwide. But he gets nothing out of it all. Another item was the **standard decorative building block** (concrete), after creating 2,000 prototypes he could still not devise a viable manufacturing process. He then used these blocks in other later projects. The **'STYROCRETE'**, **'STRAWCRETE'** and **'WOODCRETE'** products involve specialised plastics with particular insulating qualities, and are unsinkable and recyclable. Fairly new off the drawing board is a **low-speed rotating generator** delivering power comparable to present high-speed generators. Also quite recent is a new **catalytic converter** for car engines that is fully effective but employs only 10% of the Platinum required by conventional converters. This conserves a scarce and expensive element. However, it remains to be seen whether this difficult market will take up the product. Branko has enough projects to last him through the next

millennium, if only he had some financial resources! He uses an interesting expression *"the scope and nature of my creative curse!"*. However, his greatest invention may be the way he tries to blend half a dozen inventions into one patent application to keep costs down. His Patent Agent is not too crazy about it. But many of us probably do the same thing without realising it.

Mr Trevor Baylis, OBE.

A household name because of the clockwork **radio**. The prototype was available by 1992, and in 1994 a factory opened in South Africa to manufacture them. Trevor visited the factory and met with President Mandela. In 1998, he was awarded the OBE for the humanitarian aspect of the work. His previous inventions were known as **'ORANGE-AIDS'** and designed to be of assistance to disabled people. Trevor joined the Institute in 1981 and was a Council Member from 1987 to 1998. The radio was initially marketed under the name Bay-Gen. From 1996 Trevor has been on the Steering Committee for the National Innovation Centre (NIC). Trevor has for several years also been advocating an Academy of Inventors.

Mr John Bennett.

His invention in 1984 was the **'SUPERBOWL'**, a simple device. A new washing up bowl which also serves as a colander/rinser for vegetables. One side was raised with slots for drainage, produce could be left under running water with no chance of an overflow. The bowl is used either way up for different functions, and costs no more than an ordinary bowl. He is MD of Lenscap Ltd, the company marketing the product.

Mr H. H. Berry, MIEE, FInstPL. Died in 1939.

A Founder-Member of the Institute. He invented **'Safety-First'** medium pressure ironclad switchgear, which was branded as **'Foolproof'**. His **'MAGICOAL'** electric fires, with 'Chameleon' colour change effects in the fireplaces, were installed by the Royal Family and used by the Royal Navy. He also devised many labour-saving devices for the home. As a licensee, he was involved in **'continuous concrete conduit laying of cables'**.

Mr D. J. Blaikley. Died in 1930s.

One of the first Members of the Institute. A manufacturer of musical instruments, and the driving force at Boosey & Company Ltd. An innovator of instruments and toolmaking used in their manufacture. In 1878, his patent for **compensating pistons for brass instruments** revolutionised design of virtually all brass and wood instruments (except trombones). The company became internationally known, and as an expert on Acoustics he gave lectures to the Royal Military College of Music. He was sole manager of Boosey & Co until 1919, when his son Arthur took over. From then onwards Blaikley Senior restricted himself to R & D.

Dr Barrie Blake-Coleman, BSc, MPhil, Dip.BA, PhD.

A Fellow of the IPI. His qualifications are in Science and Business Administration, with a doctorate in **Metrology** and Engineering Management. His first invention was in 1970, the **'KNIGHT' anti-intruder alarm** which was licensed and on sale for 7 years (unpatented). In 1971, he devised a more aggressive device but medicolegal advice was that it contravened 'Man-Trap' laws. Throughout a career with BAT and Philips, he developed responsibility for work involving **biosensors**. From 1983 he worked at CAMR (Porton Down), managing a large group in biosensor research, many patents resulted. Barrie is now Commercial Product Development Manager with Nottingham Trent University, and also independently operates as 'Merryfield Consultants'. He has 14 patents, 7 are licensed and 4 already on the market. Barrie has been Chairman of the **Steering Committee for the National Innovation Centre (NIC)** since 1995. His initials are BBC, which causes confusion in document terminology.

Black, Jet. (see Jet Black) . Stage name.

Mr Peter Blanchard, AMIMC.

Peter is a Member of the IPI, and an Associate Member of the Inner Magic Circle (Silver Star). He markets his products under the name 'Blan-Ventions'. In 1996, in the '***INVENTORS WORLD***' magazine he published '***World of Magic***', discussing the role of inventions in the art of being a magician (but not revealing the know-how). His inventions include the **'BLINDSPOT MIRROR'** for motorists. In 1976 he wrote a thesis on the blindspot problem, then in 1982 he introduced his mirror designed to see into the blindspot. Simultaneously, he introduced a children's game called **'VOLCANO'** which had excellent sales in Europe/UK. He is often referred to as Peter 'Blindspot' Blanchard.

Mr Denys Bliss, CEng, DLC. Died in 1983.

A Member since 1973, and a Council Member from 1978. A radar mechanic in WW2, then an aeronautical engineer for De Havilland working on the Comet and other planes. From 1960 he was absorbed in concepts like hover-trains, and was Chief Engineer for Tracked Hovercraft Ltd until 1969. Denys invented the segmented base, known as the **hovercraft skirt**, whilst working with Sir Christopher Cockerell. His development is now used on hovercraft worldwide. However, from 1970 he operated his own company Bliss Pendair Ltd, concentrating on a suspended monorail principle. The **PENDAIR SYSTEM** was successfully demonstrated in the Far East.

Mr Paul L. Bone.

Trading as Lizant Innovations, based in Newport in the Isle of Wight. Two products have been successful in SE Innovation Competitions in recent years. The first was the **'Miracle Watering-Can'** for plants, the next was the **'Feed 'N' Treat'** device which fits into the spout of a watering can and dispenses pesticide or weedkiller at the rate the pellet dissolves in the passing water. The watering can design is such that water is delivered to exactly where wanted, and a push-button operated valve releases the water. It conserves water and results in stronger plants (you water the plant not the nearby weeds). The can has been endorsed by ICI Garden Products. Another invention is the **'Multi-tool paint-easy'**, a device which clips onto paint-pots and offers various convenient features for the painter. The Miracle can has now been licensed internationally. Following these successes and becoming established as an innovative company, Paul went further and set up an Isle of Wight organisation for inventors. The RTI **(Round Table of Inventors)** meet monthly, under sponsorship from Business Links and with support from SE Hants Enterprise Agency. There are already more than 20 members and this gives valuable support to these somewhat isolated individuals through the regional group and the company itself, Lizant Innovations.

Mr John Boydell.

A Member for many years, MD of Custom Doors Ltd, he was a Consulting and Forensic Investigation Engineer in cases involving personal injuries, industrial accidents, machine and plant equipment failures and faults. An expert on doors and farm machinery, living in Telford. John was admitted to the register of qualified mediators of the British Academy of Experts in 1962.

Lord Brabazon of Tara, MC, MP, GBE. Died in 1964.

John Theodore Cuthbert Moore-Brabazon. An Honorary Vice President from the early 1920s to 1960. He chaired many Institute meetings. As Lt Col. Moore-Brabazon, he co-produced the ***'ARMCHAIR SCIENCE'*** booklets (with Prof. Low). He was associated with The Hon. C. S. Rolls in ballooning and gliding adventures. He gained the first pilot's certificate and drove a car with **FLY 1** on the licence-plate. He constructed his own plane and flew it with only half a pint of petrol and wearing no boots (to minimise weight). On one occasion he took a pig on a flight ('pigs DO fly'). While Director of Photography Services in WW1 (Royal Flying Corps), he invented a useful **automatic camera**. During WW2, he was first the Minister of Transport and then of Aircraft Production. The **BRABAZON** plane was mainly his doing, and his efforts paved the way for aircraft such as the Comet.

Mr Douglas Buchanan

A recent member, the inventor and manufacturer of the '**SPECTANGLE**', a pendant for storing sunglasses when not in us, and special boots for dogs when encountering hazardous surfaces such as broken glass. His completely redesigned golf putter, the '**BUCHANAN PUTTER'** is on test with international golf stars.

Mr Andy Chipling.

A Fellow of the IPI, and Council Member until recently. The creator of a much-needed device, '***SWITCHGUARD***', designed to prevent essential electrical switches from being switched off or disconnected. This is important with medical apparatus, such as monitors or supply-machines for patients in intensive care, and for blood-banks, vaccines, veterinary and human diagnostic reagents storage equipment. Andy got the product taken up commercially and it was selling well. However, that company collapsed and he now has to find other licensees. He has amusing but scary stories of hospital cleaners switching patients off life-support systems while they vacuum (but not any more!). Andy has arranged IPI meeting venues at the Imperial College of Science and Technology where he is on the staff. He was Treasurer of the IPI in 1992.

Sir Christopher Sydney Cockerell. Died in 1999.
An Honorary Fellow and active supporter of the Institute for many years. At Cambridge he worked on radio and radar as an engineer, before turning to Hydrodynamics. In the 50s he invented the **hovercraft**, after experimenting with air as a lubricant beneath the hull of the craft. A full-scale hovercraft was built in 1958. Knighted in 1969, he should have become wealthy from the invention. He attended high-profile Institute events such as the House of Lords Reception in 1974, as an IPI representative. He was assisted in the hovercraft development by Denys Bliss (another IPI Member), who invented the skirt component so vital to the machine.

Mr Harry Cole.
Not strictly an inventor, but a holder of design registrations for board-games in the 80s. Harry was an 'ideas man', and an expert in Commercial Property Acquisitions. He realised that few inventors knew how to protect their ideas and in 1992 he published **'Ideas or Inventions can make Fortunes, How to make Yours'**. A self-help advice manual for the innocent. Harry has acted as one of the judges for events such as the Great British Inventions and Innovations Fair.

Anne and Bill Collins.
Members of the Institute. Both are music teachers. Bill was a musician in dance bands in the 20s and Anne has been a music teacher for 35 years. Bill managed the Rock band '**Badfinger**'. They co-wrote the song '***Without You***' which Nillson took to No. 1 for some time. Bill realised that he needed proper training before writing any more (he played by ear). His wife Anne taught him music and some theory, and in the process they developed a new way of teaching music with electronic teaching aids and books. They promote their products via Collinsian Concepts Ltd.

Mr Seamus Connolly.
A Member, whose invention was successful enough to get his company accolades from the CBI. His product the '**FASTANK**', a special portable water tank which is now used in famine areas. It weighs only 150 lb, holds 2,000 gallons of water and is quickly set up, or dismantled, even in difficult conditions. The tanks can also be used for small-scale fish farming. Seamus soon had orders from 55 countries.

Mr Malcolm Cook.
A Member of the Institute. The Marketing Director of Sphinx Exhibitions Ltd (based in Bournemouth). A professional exhibitions organiser who initiated The Great British Innovations & Inventions Fair in 1995 at the NEC in Birmingham. The venues since then have been The Barbican and Earls Court in London. More recently, Malcolm has also been the organiser in collaboration with the BBC 'Tomorrow's World Live' Shows. Such shows are usually in conjunction with a Business or Consumer Show, separately but at the same venue. There have been some learning experiences in the exhibiting of inventions, even for an experienced organiser like him. So in his own way Malcolm has contributed to the progress of innovation and invention, making such annual events almost routine.

Mr Cecil Cope.

In 1971 his daughter heard a radio programme on which a factory worker explained how the job of **shelling hard-boiled eggs** was horrible. She told her father, who was a design engineer in the poultry industry. He went to this factory and spent 2 days studying the problem, then designed a machine to process 8,000 eggs per hour. He was reluctant to put people out of work but realised progress was inevitable. At the age of 8 he had designed an **apple-peeler** for his mother. Early in his career he designed **poultry incubators** and **hot-water chicken-brooders**. During WW2 he designed **bomb-racks** for Lancasters, and other military devices. His first post-war invention was a **'Wildmesh'** system for making animal cages for labs, and also a **'plucking machine'** for poultry. This involved **'rubber-fingers'** which removed feathers completely, without damaging the skin (the bird's skin). He wrote a thesis on the preparation and shelling of boiled eggs (a problem even now for caterers).

Reverend Jonathan Copus.

A Council Member of the IPI. As well as being a Parson, he has been a Teacher, Broadcaster, and Journalist. Jonathan is the MD of Dentron Ltd, an innovations company. He invented an electronic antibiotic, **'THE BIOGUN'**. This is a unique new approach to antibiotic treatment of surface infections, e.g., Athlete's Foot, warts, and dental decay. It is particularly valuable in dental infections, hence the name Dentron. Jonathan has been successful in applications for innovation grants and Smart Awards. He knows the right approaches to applications for funding, so he set up another company 'Immedia TV & Radio Services', which trains young people to make the best use of the media. He is helping other inventors in all kinds of unusual ways.

Mr William Courtney, MPhil, MSc, Dip. Mar, BSc.

A Member of both the IPI and IPDC prior to the merger. He is a prolific inventor and trades as Cheshire Innovation (in Altrincham). Bill has particular interest in inventions for outdoor recreation. His more notable inventions include a **Shelter and Survival (SAS) Bag**, a **DNA double-helix dog restraint, displacement pumps** with modern novel features, **multi-lingual 'talking' information boards**, a solar compass and a shock-absorbing liquid for vehicles. The **Solar Compass** is particularly interesting, it attaches to a map and in sunlight determines orientation relevant to the map. Bill sees it as a brand promotional item. The **shock-absorbing liquid (SALi)** is intended to reduce the likelihood that when a vehicle hits a pedestrian the result will be a fatality. He has been researching this material at The University of Manchester, in collaboration with their Dept of Engineering (where he is a Research Scholar). The results suggest that his design and materials would meet millennial safety legislation for vehicles (internally and externally). He has numerous other patent applications underway.

Professor W. Cramp, DSc. Died in 1940s.

A prolific inventor, in 1922 he won the Royal Society of Arts Medal for a method of **pneumatic conveying**. Later he devised a **single-phase motor**. Other successful inventions included a method of producing **nitric acid from air**, and a new **coal-face lighting system** (as co-inventor).

Mr Henry B. Crawford.

A definite contender for the status of oldest current Member of the Institute, since he was 74 when he joined (and that was a while back). In 1999 he was an exhibitor at the Tomorrow's World Inventions Show, and had been worried that he would not be allowed to exhibit at his age. He has designed a 2-wheel trolley which carries a heavy load in an upright position. It is suitable for gas cylinders and suchlike, there is a clever mechanism whereby the load is raised slightly and seated vertically. This system makes the movement of such loads relatively effortless compared to conventional trolleys, which are moved in a sloping position.

Mr William Darby.

The 1985 winner of the Institute's Richardson Gold Medal in 1985, for his **'Fuel Pellet'**, an innovation in energy conservation. This solid fuel consists of a mixture of one or more paper products and one or more oil products, i.e., waste paper with fuel-oil/waste-oil, containing a binding agent. Shredded paper is conveyed to a feed-hopper and onto a palletising press where oil products are added via a conditioning unit. The Third World can benefit most from this conservation process.

Francis W. Davis.

An Honorary Fellow from 1968. An American, the inventor of **power-steering** for the automobile. The award was presented to him at The American Academy of Applied Sciences meeting to honour his achievements. This was an incredible invention considering that he was not in the automobile industry.

Mr Travis Jack Dawes. Died in 1971.

A Fellow since 1954. He had been Technical Engineering Consultant to the Greyhound Racing Association for about 30 years, and was latterly a Director of White City Stadium Ltd. One invention, an **auto-timer for mass X-ray photographs**, was used in hospitals throughout the UK. Following installation of the UK's first 'Totalisator' (The Tote) at Newmarket in 1927, he was permanently involved with them. Although primarily an engineer, during the war he was in Radiation Physics. At the time of his death, he was working on a **linear-motor controlled lure**, and **automated coin-operated tote tickets**.

Mr Eric P. Donald, MSc, ARAeS, TEng.

Winner of the Richardson Gold Medal in 1977, for his invention relating to **metal fatigue detection**, bolted joints and transport. He has a Masters Degree in Aeronautics, and worked with Donald Hathaway Associates. He has served as President of the High Speed Monorail Institute. His detector was known as the **'BLEEDING BOLT'**, and it received much attention from the Americans at the time of the Chicago DC10 disaster in 1979. In connection with all the work there, he emigrated to the USA.

Mr Reginald Charles Draper, FIED. Died in 1984.
A Member of the Institute, he spent his life working on steel structures and associated machinery. The designer of a variety of dock-gates, he was good with moving structures. A member of the team developing the **THAMES BARRIER**. His design for a **Rising Sector Gate** was selected. A Londoner who took great pride in the Freedom of the City bestowed on him in 1982. His own illness and death followed a family tragedy.

Mr C. O. Forestier-Walker.
A Council Member for many years. An expert in Third World Development. He was originally trained as an aeronautical engineer. Since 1944 he has managed innovation and invention in Civil Aviation and Nuclear Engineering. He developed new techniques for construction of low-cost water-retention systems. Inventor of the **Trammel water-drainage and harvesting system** for poor rural areas. Many of his ideas were adopted by the MRP Conference for Latin American, Caribbean countries & Mexico, 1982. His business is Minola Industries in Aldeburgh, Suffolk

Mr G. D. Gillies. Died in 1977.
An Honorary Member of the IPI. Formerly a Principal Examiner with the Patent Office, he represented the Institute at the long and tedious discussions of the inter-governmental congresses for the European Patent Convention. In the IFIA Bulletins, he reported on progress of the introduction of the EPC system (1970s).

Mr Sam Green, CBE, CEng, FIEE, MIProdE, FRSA. Died in 1996.
A long-serving Council Member, Chairman of the Institute for many years. He was also prominent for leading the International Federation of Inventors Associations (IFIA) throughout its formation years. A recipient of the **WIPO Gold Medal** in 1984, and also involved in the Trust Fund for the **Richardson Gold Medal** of the IPI. It is often forgotten that he was an inventor in his own right. He was associated with the electrical and textile industries for long periods prior to directing Remploy Ltd (for which he is most known). During those early years, as a draughtsman and development engineer, he invented several valuable devices, the most well-known being his **4-colour box motion for automatic looms** (1933). In 1944 he built a factory to produce **special laminated plastic sheets**, for use as fire barriers on ships torpedoed by U-boats. Another connection with innovation was from 1945 when he joined 3i's (ICFC then) as an industrial advisor.

Mr David Griew.
Winner of the Richardson Gold Medal from the Institute in 1983, for his '**MICRO-TEE RULER**'. He had several projects which did not reach completion, but learned how to present a good image to manufacturers. He decided that small drawing boards without a stand had serious deficiencies. He corrected this by designing a tee-square to go directly onto the paper. But paper is quite thin, so this did not work. The eventual result was the Microtee Ruler which did a job that had never been done before. Even now potential purchasers are sceptical at first. It has achieved commercial success.

Mr Frederick Marten Hale. Died in 1931.

A Founder Member of the Institute, and a household name during WW1. An engineer, who worked for the Nobel Dynamite Trust. At that time he started inventing weapons. The Hale **rifle-grenade** was much used in the first war (10 million). He also invented a **depth-charge**, which accounted for numerous U-boats. Hale also invented **aircraft bombs**, the one that Commodore Samson used to destroy the Zeppelin shed was made by Hale. As mentioned previously, he was unhappy with the government's treatment of his patent. He expected 5% royalty, which would have been £150,000, but ended up out of pocket. An extreme case of wartime invention being unprofitable, and the need for the Institute.

Mr Ian Hawes.

A well-known Member and frequent exhibitor, he invented the **'Honeycomb Flood Barrier'**. This is a sloping plastic honeycomb arrangement that prevents a few inches of floodwater from getting into the house. It is designed for homeowners who live in areas that frequently experience some flooding. The sludge and pollution inside buildings is prevented since the height of the honeycomb cells increases from the front to become several inches high at the back (which is next to the building/door). Water entering the honeycomb reaches a level which results in a water gradient which then reverses the flow of water so it runs back out.

Stephen Henecker.

Techmart Inventor of the Year in 1985, for development of the **'Touch-control tap'**. Actually two of them set out to create the product, but his partner Leslie Reeve died in 1982. Later, Stephen went on holiday taking a prototype with him. In the garden of the holiday cottage, the prototype worked. He built a cut-away wooden illustration model, and filed a patent application. There were many delays in getting design approval. By 1985 he was giving up. In ***'THE INVENTOR'*** magazine he saw details of the Techmart Competition and entered as a last resort. He is now a believer in the 10 golden rules for inventors advocated by Bob Symes.

Mr Ron Hickman, OBE, MISA.

A Member since 1973, now a Fellow and former Council Member of the Institute. The most prominent Member in recent times, he was South African by birth, coming to England in 1955, aged 22, becoming a British citizen in 1962. Initially working at Ford, he later joined the founder of Lotus cars, Colin Chapman, dealing at first with production problems on the 'Elite'. By 1965 he was Design Director, and had complete design responsibility for the original 'Elan' introducing novel ideas such as **body-contour bumper panels, retractable headlights**, and **mounting individual registration numbers on a mesh radiator grill**. Ron purchased a new bungalow and did a lot of DIY for which his Lotus mechanics knocked up a crude, barely portable, sawhorse with a welded tubular frame. A worktop was formed from two spaced wide, but fixed, beams with a 'Record' vice at the front. It was so useful that Ron began to glimpse commercial prospects, and the seeds were sown for what was to become the **'WORKMATE'** ®, but a saga more fraught with problems than we might guess followed. Several design stages followed, but if the original top of his first unit had been a single board, rather than two spaced fixed beams, his subsequent 'Eureka' idea - that one of these beams might move to form a huge twin-screw vice- would not have occurred to him, and a commercially viable design would never have seen the light of day. Despite Ron's numerous presentations of his final design to likely manufacturers, including Black & Decker, the commercial potential was unrecognised by all of them - Stanley even estimated (in a letter dated 1st April) that the 'potential was in dozens rather than hundreds!'. In 1968 Ron tooled up to produce **'WORKMATE'** himself in an old brewery, after his Patent Agent ***Mr Mike Roos*** (also an IPI Member) commenced protecting all aspects of his bench worldwide. In 1972, after Ron had sold 25,000 benches by mail order, Black & Decker recognised the potential and sought worldwide licences. To their delight they discovered that Ron had just finished designing a lightweight dual-height model. Black & Decker introduced this product, with national TV marketing. Sales took off and they just could not keep pace with the exponential demand. By 1982 sales had topped 14 million units, and are now well over 50 million. Ron and Mike Roos outlined this extraordinary story in a paper read to the Chartered Institute of Patent Agents in 1982, by which time over twenty successful cases of Patent infringement had been concluded in all parts of the world, necessitating huge amounts of travelling for both Ron and Mike to brief lawyers, and attend Court cases at which Ron gave crucial evidence.

For some while Ron ran a development company in Jersey, in an attempt to find new ideas, but 'lightning did not strike twice' and, after investing huge sums without much success, it was disbanded some years ago. Ron now lives in Jersey in a remarkable 'futuristic' house, which, as one would expect, is full of fascinating and immensely practical gadgetry. Ron's OBE was awarded in 1994 'for services to industrial innovation', (believed to be the first time such a citation had been used). This fascinating story continues to attract attention around the world. In the last few years it has formed the basis for a design course run by the Open University, and legal aspects of the story are regularly examined by student members of CIPA.

Mr Robert Hobbs.

The designer of an ingenious cat-flap that doses the cat with flea repellent liquid, as the cat passes through it. This concept is meant to ensure that the cat gets treatment at known intervals. The ordinary flea collar suffers from the fact that you never really know when it has become ineffective. His description of it is a **'dual purpose cat-flap'**, which 'lets the cat come in but keeps the fleas out'. Great news for cats, and bad news for fleas. It is battery-driven when switched on. Robert trades as Combined Flair Ltd (in Devizes).

Mr Stephen Hobday.

The designer of an ergonomic type of keyboard, trading as PDC Maltron. Hand movements are much easier and users report recovery from Repetitive Strain Injury and similar problems. Stress-free operation results from use of the range of keyboards which are available for a variety of occupations. One type is specific for postal sorters and has a one-handed operation (left-handed) whilst the worker has the postal item in the other hand. It is called the **'MALTRON KEYBOARD'**.

Mr Fred H. Howorth, FRSA, FIIC.

A Member from 1970, and a Fellow from 1974. He is an expert in Air Technology and designed a **sterile operating theatre area**. His Fellowship was gained by ubmission of a thesis. Previous operating room history involved a 4% infection rate of patients from viruses or bacteria. In his design, patient and staff are enclosed in a transparent compartment through which sterile air is exchanged at the rate of 300 times per hour. Exhaled air from patient and staff is continually removed, it cannot contaminate air in the enclosure. The introduction of clean air and removal of contaminated air is achieved using silent-running fans, and high-efficiency bacterial filters and diffusion units. Such systems are now commonplace worldwide.

Mr Martin Hutchinson, FRIBA. Died in 1974.

A Member from 1967. He was an architect involved in building production plants for Ford motor Co, and had also been involved in construction of the prefabricated concrete harbours (Mulberry Harbours) used in WW2. A patentee of designs for **road viaducts, vertical clocks** and other unrelated systems. He studied the building principles of the Ancient Greeks. He **painted watercolours of buildings due for demolition**, and has many **model ships** in the National Maritime Museum, He was well known at the Institute, since his offices were in Abbey House as well.

The Right Hon. Lord Ironside, Hon. FCGI.

Edmund Oslac Ironside, 2nd Baron of Archangel & Ironside from 1941. The son of Field Marshal Ironside. A historian who has published diaries available through the family. A Vice President of the Institute from 1977-1991. Lord Ironside has a very high technology background, and in the government has played a leading role in Defence and Scientific policy-making. From 1984 he was Vice President of the Parliamentary Scientific Committee. He is currently Chairman of the Advisory Committee for the Science Reference Library (British Library). Chairman of an all-party Defence Study Group from 1995. From 1989-1995, he was a Defence Consultant with Rolls Royce Power group. He lives in Colchester.

Jet Black,

A recording artist for the past 25 years, drummer with the internationally known group '***The Stranglers***'. Jet (stage name) is a Director of several companies in the music industry. For about 20 years he has been concerned with the way that the large bass drum dictates the location of the drummer in the band/group and how it restricts some of the drummer's activities. 5-6 years ago he found an answer, then he joined the Institute and got advice and support from several members. ***Brian Wybrow*** helped with prototypes, ***Pat Russell-Rayner*** with patenting. After almost 15 years development, the **'Jet Black power bass-drum pedal'** product is the result. The electronics enable the drum to be played from any (remote) location. Also several drums may be activated simultaneously by one foot movement. The drumming may be superimposed live onto pre-programmed music. Drummers are impatient to get one of the devices: drummers everywhere have waited for such an invention. In the New Year, it will be launched on the market. Few of the members probably realise that even a drummer might be an inventor. He will do well.

Mr Barry Jones.

Barry Jones and his partner Nabeel Shirazee invented the **Magnetic Suspension** device that attracts so much attention. In 1996, they were awarded Gold Medals at Geneva and Morocco exhibitions, and at INPEX in the USA they had Gold plus the Grand Prix. Following some excellent publicity from ***'INVENTORS WORLD'*** they never looked back. Barry says that by following the advice of the IPDC (now IPI) to exhibit at INPEX, they were a year ahead in development. Using innovative electronics their suspended globe is eye-catching and thought provoking. In a shop window the device cannot be ignored. They have a contract with an American company now and commercial success is guaranteed. There were so many enquiries that they set up a new R&D unit in Cardiff to cope with all the applications.

Mr David Jones.

In 1973, he was a dental student. But intrigued with a 'spheres within spheres' device he was messing about with, David suddenly conceived a clock which could be seen from all directions. He developed this concept of a double-helix visible from wherever one was in the room. The **'OMNICLOCK'** was the result. It was extremely modern and stylishly coloured. The cylindrical device could be placed vertically or horizontally and the spiral numbers indicating the time were always visible. The Queen Mother owns two of them.

Mr Mike Keenan.

Mike has invented a material which can be used to **fireproof almost anything**, it can be smeared on as a coating. There were 2 events that featured in his realisation of this unique product. Firstly, he had been working with his brother as a painter at Tilbury Power Station in the 70s, they observed that mixing some pot-ash lying about there with their paints resulted in the mixing-stick becoming fire-resistant. Over many years he investigated different **materials for fire-resistance**. He was able to make newspapers and **cardboard non-flammable**, later he had success with plastics. An important aspect was that **heat did not pass through the plastic sheet**. This was a valuable attribute. But the second phase began after the King's Cross underground station fire. Mike gave much thought to the fact that the fire was fuelled by the lubricating grease on escalator wheels and runners. This set him on the trail of ordinary **grease with good lubrication but fire-resistance too**. Following some publicity, Mike has had help from Business Link Hampshire, they formed a virtual company for him (Keenan & Associates, in Dorchester). A team of experts is now involved and they gained a SMART award from the DTI, so things are moving fast. One spectacular bit of publicity was in 1998, at the ceremony for the IPI/IPDC merger, Mike smeared Howard Stableford's hand with the new grease and then attempted to light it with a propane gas torch capable of

1,000° Celsius. Howard is still in one piece, we await the outcome of all this spectacular development to see what happens in the end commercially. Late news is that Keenan & Associates got the runner's up award at the 1999 South-East Innovation Competition in October (Business Category).

Mr Hugh Knight.

In 1960 he patented and started making a loft-ladder called **'FOLD-A-WAY'**. But it took 7 years before a little factory was set-up, then he was invited to submit to the Design Council for registration. The ladder has been modified and improved several times since. At the Brussels International Exhibition he was awarded Gold and Silver Medals for two different versions. Later there was the **windowsill fire-escape ladder, the gantry escape ladder, surveyor's ladder** and **yacht ladder**. In 1968, he introduced the **'Uni-folda' ladder**, a multi-purpose product. At LINPEX he also showed a **remote bait dispenser** and various fire-escape ladders. Hugh was a precision engineer by trade, from 1955 he had his own company and the inventions stem from that date.

Mr Peter Knight, BSc, CEng, MBCS.

A Fellow since 1998 and a Council Member in 1999. Mathematics graduate, whose early employment was in Data Processing, Information Technology and Logistics Management. Whilst working for the UK head office of Omega, Tissot and Swatch watches, he won a British Computer Society award for his development of a **voucher system** for fast-moving consumer goods (FMCG). At the 1999 INPEX Exhibition in Pittsburgh, he received an 'Outstanding Achievement Award'. He is Research Director for ***'INVENTORS WORLD'*** magazine, and he has functioned as a judge in several International exhibitions. He is an inventor and has some patents pending. **'MAGIC MIST'** is a plant growth formulation that prevents wilting in hot weather, so plant grows at maximum capacity throughout. The **'Universal Mapping Guide'**, is a new approach to getting one's bearings in a strange location. Peter is also a Director of ACTI (GB) Ltd and involved in production of directories of inventors and reference books. One major project underway is the **'PRONTOSAURUS'**; a lexical word-mapping system.

Mr Kane Kramer.

A Council Member of the Institute. He is from a family in the footwear/clothing trade and he progressed into selling cowboy boots, but then went into promotional aspects of entertainment. By 1979 he had finished 10 years development work on a credit-card sized digital recording and playback device (all solid-state). He had intended to use the name **'IXI'** in relation to the product. This project did not get anywhere, and many years on he was in Hitchin doing exclusive kitchen design work. One project required matching colours for all appliances, but there was nothing appropriate on the market. In 1995 he created a metallic paint that looked the part. The specification was for a matching colour for cooker hoods and grills, but the product had to be water-based, non-toxic and metallic looking. He filed for a UK patent, then lost a year in negotiation with a major UK company. Getting private backing for the international patenting was difficult (£30,000). Then he exhibited overseas (including Malaysia). By January 1997 he was ready despite having to find £60,000 more finance. His product **'METALCOAT'** was exhibited at the Barbican and to his amazement he won the '*Sunday Times*' Inventor of the Year Award for 1997. Now he is in production, the company is called Metal Paint Ltd, using the **'IXI'** trade name. Already he has a joint-venture company based in Beverley Hills (USA). The future looks bright (and extremely metallic). He has other spin-off ideas surfacing already.

Mr Paul Kuehnrich. Died in 1932.

An inventor, steel magnate and benefactor of Sheffield (his adopted city). Born in Saxony and being tall was thought older than he was. At age 16 he was selling steel, at 17 he emigrated to England, became naturalised and set up in Sheffield as a metallurgist. He became associated with Darwins Ltd and introduced mass production of **razor blades** in 1924. The process was from raw material to finished product. Darwins became a group of companies, and Paul shaved the World but never had a shave himself. During WW1 he invented a substitute for Tungsten Steel (which was scarce because of military usage), it was **Cobalt Chrome Steel**. Later he created **'STAINLESS STEEL'** and then made **'CRYSTEEL CUTLERY'**. His final business was called the 'Universal Rustless Steel Corporation Ltd', which produced razor blades with a super-keen long-lasting edge. He had some difficulties raising finance in later years, but created business for the country and much employment in Sheffield. He committed suicide aged of 77 (nothing left to do).

Mr John Laing.

Inventor of the **portable walking-stick seat**. It transformed instantly into a tripod seat and so did not require (like previous types) the user's legs to be part of the support. The stick portion is pressed into the ground, the tripod opens, it weighs the same as an umbrella. About 1930 style.

Professor Eric Laithwaite, DSc, PhD, CEng, FIEE. Died in 1997.

The author of a book on how to invent. Associated with the Imperial College of Science and Technology in London from 1964. His field was Heavy Electrical Engineering. A prolific inventor, many items concerned with **linear motors**. Very interested in gyroscopes and well known for his involvement with 'Magnetic River' **monorail high-speed transport systems**. In later years he did Consultant work. He claims he would rather invent the problems than the solutions. He has published many articles in ***'THE INVENTOR'*** and presented papers at Institute meetings.

Mr Barry Langrish.

Inventor of **'WINE-STEMS'**, a set of unique shaped rods with attachments at the top to hold wine bottles or wine glasses. The rods are stuck into the ground, and result in drink containers all being at hand height when seated. The system is for outdoor use at garden/beach parties. It has been commended in exhibitions (1998), and Barry now has a manufacturer committed to producing and distributing the products in the USA and UK.

<u>Sir Harry Legge-Bourke, KBE, MP.</u> Died in 1973.
An Honorary Vice President from 1966. Since 1945 he had been the MP for Ely. In WW2 he was a Guards Officer, ADC (Major) to the British Ambassador in Cairo. His parliamentary career involved Defence and Foreign Policy, and also Science and Technology. He eventually succumbed to deafness, otherwise as Chairman of the 1922 Committee he would likely have become the Speaker of the House of Commons. In 1969 he was largely responsible for linking the NRDC with the Institute's first post-war London Inventions and New Products Exhibition (LINPEX). Shortly before his hospital appointment for an unsuccessful operation, he headed the Institute's representation to NRDC in 1973 with a view to expanding the role of the Institute in commercial promotion of inventions. He did not survive the operation and so never learned of their success.

<u>Mr Stan Lewis & Mr John Sandham</u>.
Two officers in the London Fire Brigade, Members of the IPI from 1995. Joint inventors of **'STAiRMATE'**, a small and portable platform that enables an ordinary ladder to safely reach any area above a staircase. It was inspired by emergency and rescue potential for obvious reasons, but plainly has many everyday uses in the home and industry. It is manufactured under licence by Easy-Dec Access Systems Ltd. The product was recently granted **Millennium Product Status**, which is a great boost for the 2 inventors, and will help to assure the product of commercial success. They have other projects underway, and have formed their own company, 'Lewis Sandham Concepts', in order to facilitate further inventions.

<u>Mr Alan Lovegreen.</u>
A former Council Member of the IPI from 1974, a member of the Finance and Planning Committee. He was from a family of inventors. In 1975 he had been making money from inventions for about a decade. In 1964 he had conceived a **de-watering conveyor belt** for sand and gravel. The machines were modified to cater for manure, pig manure, protein, sewage. His machines brought in licensing prospects from round the World.

<u>Mr Frederick W. Mackenzie.</u>
A Member for many years, Fred invented **'LETRASET'**® the 'do-it-yourself' type transfer used in artwork functions world-wide A Fellow of the Institute of Printing, having been the Founder of the Association of Printing Technologists. His involvement with printing and graphic reproduction was lifelong. During WW2 he was involved in map production, in psychological warfare (India), and later in a printing operation in Calcutta producing documents for military intelligence. Having the concept of Letraset, he assembled a team of experts in the various activities needed to make and market the products. It was initially a wet process which gave a permanent image. However, in the USA there was insistence on a dry process. So the ultimate product marketed was a dry process which did not produce such permanent images as originally intended. But it was a great success. In later years he was associated with the Inventor's Workshop, at Paddington College.

Mr Nicholas Mark.

One of the newest members of the IPI. An American living in the UK, he invented the **'K-MAT' (knee mouse-mat)**. This is a curved mouse mat that clamps on the thigh and minimises repetitive strain disorders, because the mouse is used without the arm extended. It is excellent for use when in an armchair. In October 1999, he won the award for 4th place in the South-East Innovations Competition (Individual Category). 'K-mat' has not yet been advertised directly, but over 500 units have been sold as a result of media reviews following major exhibitions. It will undoubtedly be a success. He now lives in Newcastle and trades as Nicholas Mark Innovations. He has a mass of letters from grateful customers already. Nicholas is preparing for mass production in the UK and is seeking distributors, he intends to licence out the USA production and marketing. At the SE Innovation Competition, an awards judge bought 2 of the mats, a TV Producer filming him bought one, and the owners of the Kingston Guest House that he stayed in bought one too, not to mention lots of people attending the show. It is now on trial with W. H. Smith's. He must be doing things right, which is unusual for an inventor.

Mr Andrew E. McGuire.

A Fellow of the IPI from 1980. Winner of the Richardson Gold Medal in 1979, for his **'Acoustic atomisation of fuel in the internal combustion engine'**. A complex but valuable invention taken up by the industry. This P.E.D. control device saves petrol and reduces emission pollution.

Mr Joe Michael.

A Fellow of the IPI. A computing engineer with an invention that involves 'fractals', a mathematical concept few of us grasp completely. He designed a system of **'shape-changing robots'**, which has won him many awards including the 1995 'Against the Odds' award from ***'INVENTORS WORLD'***, a Gold Medal at INPEX 1995 (USA) and The Monaco Grand Prix du Jury Award in 1996. His system permits a set of similar blocks to be computer-controlled so that every one of the blocks knows what they are all doing. The object is to perform some function from a distance by computer signals to the blocks which then align appropriately. There are many applications (bomb-disposal from a safe distance, space-walk engineering, work in a small confined space that a worker cannot get into). He describes his shape-changers as hardware doing the job of software. Unfortunately to make progress beyond this stage requires massive funding, which may come from a large company in the USA, or from NASA or similar. He has proved it works, and he has a prototype. His company is E. C. C. Ltd in London.

Mr Albert Midgley. Died in 1961.

A prolific inventor, whose life's work was exhibited at Brunel University in 1985. He had **212 patents** in the UK alone, and he was the first to create a small dynamo capable of functioning from a motor car. He was 79 when he died. Motor car lighting and starting, automatic transmission, magnetos for aircraft, automatic firing, gyroscopes, signalling lamps, range-finders, radio, loudspeakers, hi-fidelity amplifiers, sound recording equipment, pipe organs, electronic organs, bombs and fuses, automatic time-switches for appliances, traffic indicators for motor cars, the list is enormous.

Sir William Mills. Died in 1932.

A marine engineer in early career, then worked with alloys and took out several patents. He started the first aluminium foundry in the UK. In 1915 he patented the **Mills hand-grenade**, which was a big improvement on an earlier Belgian design. 7,500,000 of these grenades were used in WW1. He founded the Mills Munitions Company to manufacture them. He had other inventions, but was a household name for the grenade. Following the war, he received a knighthood in 1922 and eventually got £27,750 royalty from the government. However he was dissatisfied with the payment and fought a battle over the income tax involved. He finally lost this in 1929 (a close parallel with Marten Hale's outcome).

Mr Derek Myhill, MSc, FRSH.

A Fellow since 1998. A medical biochemist, previously Director of large Pathology labs in the USA. Also a Director of Research in Pharmaceutical companies. Creator of the **'DIFFUGEN'®** range of **immunodiffusion kits** for measurement of Immunoglobulins A, G and M in blood, products used in hospital labs worldwide for 20 years. He also created the first commercially available immunodiffusion kits for diagnosis of 'Farmer's Lung, Bird-Fancier's disease, and Candidiasis. MD of Sigma Games Ltd, which in 1990 introduced the board game **'GOLF INTERNATIONAL'**. Many inventions of a diverse nature since the 80s. **'GRAV-CHECK'** enables a winemaker to check the gravity without opening the demi-john. **'HYDROGEL'** hot/cold ice-pack (refrigerated NOT frozen), the only such product using discontinuous gel, commercially successful. **'MINICULTURE'** plant growth system involving transparent soil containing nutrients and indicator systems so that soil colour changes indicate need for nutrient or water, and with local colour changes at roots. An ideal product for schools and hospitals, the system allows starting growth at any time of year (Second prize in the 1993 IPDC Inventor's Fair). Derek is Editorial Director of ACTI (GB) Ltd, with various reference books coming down the line, including one on **WOMEN INVENTORS**, and **'THE A-Z of INVENTORS'** (as serialised in ***'INVENTORS WORLD'*** magazine). In the past 20 years Derek has aided lots of inventors as Patent Consultant and mentor. The founding of the IPDC in 1991 was due to the large group of inventors relying on Derek for help. His main project is now the **'PRONTOSAURUS'** lexical word-mapping system, patent pending, (which is inventive enough to be covered by a Patent as well as the obvious copyright), Derek is the Principal inventor on this very long-term project (7 years input so far).

Mr Douglas W. Nield, FID.

A Member of the Institute from 1967, a Life-Member as of 1972, and a Fellow since 1994. Also a fellow of The Institute of Directors. A prolific inventor with quite a few patented products on the market worldwide. His father had a valuable patent in 1920 (3-legged wheel and pulley drawer for removing heavy components from machinery). Douglas is a Founder Member of Nield's Patents Ltd, a factory in Cardiff. He patented the first **unbreakable insulators for electric fencing**. Also **steel fencing posts** which permit fast removal of netting, for farmers doing quick strip grazing. Other successful products have been a **drifter apparatus** for detecting marine currents, and the **'CADDYMATIC'** tea dispenser which ejects a teaspoonful of tea at the press of a button. At age 67 he qualified as a helicopter pilot, and has flown with the World helicopter champions in Russia and USA.

Dr K. D. Phillips

A Member of the Institute. He invented the **Phillips Helmet** which simulates the natural protective mechanisms of the human head. The helmet is designed for cyclists and has been shown by the Transport Research Laboratory to give 60% more protection than conventional helmets. His invention won the 1999 South East Innovation Competition (individual category).

Mr Harry Jeynes Pickles. Died in 1971.

A Member from 1949. He died aged 80 after many years in poor health, but was still working on an invention. He designed a **valve-balancing gear** for the steel industry, which was used all over the world and devised special **cranes for erection of huge gas-holding tanks**. Another well-known invention was a **turning device for hot steel bars** in rolling mills.

Sir Alistair Pilkington. Died in 1995.

The inventor of the ***floating glass process***, developed at Pilkington's Glass Works. The process involved melting the glass and getting bubbles out, then rolling the molten glass over a bath of molten tin. This resulted in a flat pane, which was drawn over the tin and cooled as it travelled. Perfect glass resulted, thus solving the age-old problem of bubbles on the glass surface. Sir Alistair explained the process in detail in a presentation at the 50th Anniversary Conference of the Institute in 1969. Sir Alistair had joined the family glass business soon after his higher education. He quickly saw an answer to the problems, but they took several years to perfect (it was touch and go, with a lot of investment). In 1955 he solved it and was a Director from that year, and got a knighthood too. 1952 may be regarded as the year of the invention. He became President of Pilkington's in 1985.

Professor Lord Porter of Luddenham, MSc, PhD, ScD, FRS.

George Porter, an Honorary Fellow of the IPI. Director of the Royal Institution from 1966-85. He is a physical chemist who worked on radar as a naval officer in WW2. After the war, in Cambridge, Professor Porter studied **fast reactions in gases**. Awarded the **Nobel Prize** for Chemistry in 1967. President of the Royal Society from 1985-90. He continues to work on ultra-fast chemical reactions.

Mr Colin Leslie Poynder.

Designer of a **'Flexiclip'** which has remarkable holding power. The principal use demonstrated is as a dog-restraint. The suction pad containing the flexiclip is attached to a wall or immovable object. The flexiclip design is such that the more the animal pulls on it, the clip and pad only become firmer. However, a simple movement releases the clip and the pad then lifts off the wall. Colin trades as the 'Stayput Company'.

Major H. A. Proctor, MA, MP, KC.
A Council Member in the 40s, he was elected MP for Accrington in 1941. He had been an inventor himself in the cotton and coal mining industries (as a result of living in the USA and Australia). After some years in the Australian Army, he came to England and qualified as a barrister almost simultaneously with his political career starting. He actively represented the Institute's interests in Parliament for many years.

Mr Edward Prosser.
A Member of the Institute. He invented the **'Paint and Grain'** process, as adopted by Ronseal to become their most successful product ever. Using this product any surface can be transformed into an apparently genuine wood surface.

Mr Derek Henry Prowse.
A Member since 1972, and awarded a Fellowship in 1974. He submitted a thesis on his invention concerning **flexible abrasive disks**.

Mr George Frederick Raper, MA. Died in 1973.
A Member from 1955, inventor of the **'RAPER AUTOLEVELLER'**. A major innovation in the textile industry. He was the son of a Bradford mill owner, and he invented the machine just after WW2. It was meant to simplify Worsted spinning, and his principle was adopted worldwide. The device solved many of the combers problems, but it had taken 5 years development. Among other awards he won the Institute's Hoffmann-Wood Gold Medal for Invention in 1955. Also he became an Honorary Lecturer in Textile Engineering at Leeds University. At the time of death he was working on a shuttleless loom and a process for reducing the sliver to the yarn stage, in order to eliminate 8 steps in the process. He was then 64 years of age.

Mr Walter Reid.
A prominent Member of the Institute, present at the first public meeting in 1919. He was the inventor of a type of smokeless gunpowder. He does not appear to have been involved as a Council Member but was President of the Institute of Inventors (Manchester) which was formed about 6 years prior to this Institute, but which did not exist after WW2.

Mr Arthur Whittaker Richardson, OBE, FInstPet, FCS, FRSA. Died in 1984.
A past Chairman of the Council of the IPI. He was the inventor of various devices in the petroleum industry. The one he was most proud of was a **safe oil heater** for the home or office. He was made a Fellow in recognition of his services to the Institute and his highly regarded inventions. The oil-heater won him the Institute's Yorkshire Medal. He was the Principal Founder of the Richardson Trust Fund set up to provide a Gold Medal for the best invention of the year submitted to the Institute. It was known as **The Richardson Gold Medal** but the Trust Fund was instituted by him with the financial support of ***Sam Green, Leonard Cotterell*** and ***Leslie Burgin*** (the original **Trustees**).

Mr Leonard Robinson.

A Fellow of the Institute from 1986, from Crawley, he has many inventions concerning m**icroprocessors**. Other high-technology inventions have related to **alternative energy resources**, solar energy in particular.

Mr Harry Ross, MRPS.

A long-serving member of the Institute, a Fellow from 1986, a Member of the Royal Photographic Society. Many successful inventions in the field of **photography**.

Mr. John Sandham. (see Lewis and Sandham).

Mr W. H. Siddle.

A Gold Medal winner at the Institute's exhibition in 1938, for a most needed invention, a **wire-joining tool** which had been long awaited by industry. It comprised a casing with a central housing for a worm wheel, a co-axially arranged spur wheel and a ratchet wheel. The components are rotatable and mounted on two different shafts. The mechanism of action cannot be described in just a few words. But this invention was very necessary, it was essentially for small multiple sets of wires such as in telephone lines.

Mr Timothy Snape.

An Institute Member who won the 'Inventor of the Year' Award at the 1999 BBC Tomorrow's World Live International Invention Fair. Tim is MD of Abbotsbury Software Ltd (trading as West Dorset Internet). The company specialises in all aspects of computer technology and for 20 years has been problem-solving for businesses worldwide. The **'Portland Horn'** TM, is an air-powered horn that works under water. Initially it was a communications device for scuba-divers, it is now in use in the oil exploration industry. Designing such a device for underwater is complex, his design does not permit exhaust air to escape since bubbles from the air would interfere with noise made by the horn. Instead, air is enclosed and then exhausted through an escape valve. The horn is also of interest to seismologists.

Mr Peter L. Steer, OBE, FIMgt, FIM.

A Fellow of the IPI. Awarded the OBE in 1986 for services to Ostomy and the medical industry. A prolific and successful inventor, particularly in medical prostheses applications. In 1979 he entered the Richardson Gold Medal Competition and received a special mention for his **'SURGICARE'** invention, which brought relief from pain to colostomy sufferers. He has 300 patents.

Mr Douglas Stewart.

A Member, he invented **'VISIRAIL'** a high-visibility pedestrian barrier. Douglas designed it in 1977 and it is now in production in the UK and Norway. It won the Richardson Gold Medal in 1980. Lack of visibility through barriers separating traffic lanes at junctions was causing accidents. Pedestrian barriers had been assumed to be beneficial, but they were creating problems for motorists. A slight off-set of each row of bars in the barriers gives visibility of traffic coming the other way, which is important at traffic lights. Erection of pedestrian barriers caused only 13% reduction in pedestrian casualties, whereas erection of these new barriers caused a drop in the number of vehicle incidents to 46%.

Mr J. S. S. Stewart, MB, ChB, FRCS.

Inventor of several medical devices (he is a surgeon). Winner of the 1969 Richardson Gold Medal. A Member of the Council in the early 70s, he was also Chairman of the Northern Region of the IPI and took responsibility for an exhibition held there. It would have been unusual for a surgeon to be organiser of a large regional inventions exhibition.

Mr Godfrey Stibbe.

A Founder Member of the Institute. The Founder and Chairman of the Leicester-based hosiery-machine builders and fabric manufacturers, Stibbe & Company Ltd. A member of the Textile Trades Advisory Committee from 1918-31, he insisted on modernisation of the hosiery plant at Leicester College of Technology, thereby keeping the city at the forefront of hosiery technology.

Mr Neil Summers.

Neil suffers from Ankylosing Spondylitis, and so was medically discharged from the Royal Marines. But, not accepting that Physiotherapy and drugs could be the only answer, he took up Physiology himself and studied at a college in Massachusetts. A relative who was a spinal rehabilitation expert also advised him on some aspects. For himself and other sufferers he designed the **'BACK-STRETCHER'** (alias the **'Backrack'**). It has ribs, in a wooden frame, that contact the muscles on each side of the spine. The invention tackles a big problem that millions suffer from. The judges at the 1995 'Great British Inventor's Fair' certainly thought he had the answer, because he got 'The Inventor of the Year' award and a cash prize of £10,000. He went on to get a gold medal at INPEX (USA), in the health & fitness category. There is massive interest in the concept. He has now expanded the range to include the **'Rola'** stretcher, **'Banana Crunch'**, **'Mini-Rola'** stretcher and **'Giant-Rola'** stretcher.

Mr Tim Sumner. Died in 1981.

A Member for more than 30 years. An engineer whose career turned into weapons design from WW2. Firstly designing **mines**, then sea and air trials of underwater weapons (as part of **'Churchill's Toy-Shop'** group). His designs for **anti-submarine mortars** led to an appointment as engineer in the Admiralty Production Pool in 1951. In 1969, he became Deputy Director of Standardisation, at The Ministry of Defence. By 1981 he was Deputy Director of Weapons Production at Ensleigh, near Bath. His personal inventions were items such as a **hydraulic valve** and **peristaltic pump**. The diaphragm valve continues to enjoy sales worldwide.

Mr Robert A. von Symes-Schutzmann, Comp. R.AeS.

Born in Austria, his father was a Director of Public Prosecutions who was shot by the Nazis. So when called up for military service Bob went walkabouts instead. His education includes Engineering, his main inventions were an electronic **switching device** (as co-inventor) and **'EXHATOR'**, an air sparge pipe in lavatories to facilitate ventilation. He was involved with the 'Tomorrow's World' TV show for some years. Bob is a former President of the IPI, one of his best proposals has been the **10-point questionnaire** that every inventor should periodically test themselves with. It has proved valuable to several of the inventors listed here.

Mr Ernest T. J. Tapp, MBE, CEng, BIMechE, FIAgrE. Died in 1974.
He was a Member in the 1920s, then rejoined in 1966. A Londoner, he founded Market Transport Ltd, with the help of his brother. A decade on he founded County Commercial Cars Ltd, in order to exploit his own patented designs for **6-wheel vehicle transmissions and suspensions**. When he died the business was a large operation. His MBE was in connection with his organisational efforts in formation of the Home Guard during the war.

Mr Eric Hardman Taylor.
A Member since 1951, and a Fellow for 7 years at the time of his death. He had 80 patents, as diverse as parachute harnesses, Braille typewriters, tide gauges and electric microswitches. An engineer, he founded Otter Controls Ltd which still flourishes, making **bimetallic microswitches** based on his patents. The firm is now run by his son (also a Fellow). Baxter, Woodhouse and Taylor Ltd took on many of his inventions and marketed them. Notable examples are the **electrically heated suits** and **Taylor Helmet pressure-suits** for fliers.

Lord Arthur William Tedder of Glenguin, CB, KCG, GCB, MSc. Died in 1967.
An Honorary Fellow of the Institute, along with Cockerell and Wallis. During WW2 he was Director of R & D at the Air Ministry. Later he was Commander in Chief of Middle East RAF (1943). Deputy Supreme Commander (under Eisenhower) of the Expeditionary Force, 1943-5. The British Air Marshal from 1945, and made a Baron in 1946. Being a science academic in later civilian life in Glasgow, he did not attend meetings frequently but was a judge in the more important exhibitions. Chancellor of Cambridge University in 1950. Chairman of Standard Motor Company from 1954-60.

Professor Meredith Thring, ScD, FIMechE, FIEE, MIChemE, FInstF, FRAeS.
A Fellow and long-serving Council Member, up until 1981. He resigned when his retirement to Suffolk made future attendance difficult. He had been at Queen Mary College in London, as Head of the Dept of Mechanical Engineering. His particular interest in later years was **'Inventions for the Poor and Needy'**. Well known at the Institute having published many articles on invention and having chaired Institute meetings. In 1973 he published ***'MAN, MACHINES & TOMORROW'***.

Arnold Tickner, FCIPA.
A Council Member from 1955-1964. A Patent Agent practising in London and Newcastle from 1938. Just prior to his death in France at the age of 52, he had aided the Institute in a particularly difficult situation. During the war he was in armaments research.

Mr P. C. Vincent. Died in 1979.
A Fellow of the Institute. Creator of the **Vincent motor cycles** of great fame.

Sir Barnes Neville Wallis, CBE, FRS. Died in 1979.
Aeronautical engineer and inventor, one of several Honorary Fellows in that era who actively supported the Institute. He was a marine engineer, and involved in design of the **R100 Airship** at Vickers. His particularly notable successes were the **Wellington Bomber** design, the bombs that destroyed the 'Tirpitz' warship and the famous **'Bouncing Bombs'** used by the dambusters. He also designed a 'Duralumin' bridge, a submarine modification and a large radio-telescope. From 1945-1971 Sir Barnes was Chief of Aeronautical Research for British Aircraft Corporation (at Weybridge). He designed the first **swept-wing aircraft** in the 50s. He was knighted in 1968. He lectured at Institute meetings, was sometimes a judge, and attended important meetings promoting the cause. A Freeman of London and holder of 6 honorary doctorates.

Mr David Wardell.
A Fellow of the Institute from 1998. A Director of the **IPDC** (Intellectual Property Development Confederation) from 1993, along with Peter Knight. The organisation merged with the IPI in 1998. David is now a Council Member of the Institute. From a simple IPDC newsletter called ***'THE INVENTREPRENEUR'***, they graduated to launching the ***'INVENTORS WORLD'*** magazine onto the news-stands in 1995. David is Editor in Chief. This magazine has been on sale in many countries around the World. In recognition of the success this publication enjoyed, David has much contact with the media. He functions as a member of judging panels at inventions exhibitions in Europe and the USA. He has been President of the **INPEX** judging panel in Pittsburgh for some years. In July 1998 he was appointed a Trustee of **NESTA** (National Endowment for Science, Technology and the Arts). He has aided numerous inventors by ensuring they get recognition and publicity through the magazine. In October 1999, he took on a new role as Presenter of the UK HORIZON TV programme **'TOMORROW'S WORLD PLUS'** (a cable-TV version of BBC 'TOMORROW'S WORLD').

Dr L. L. Ware, OBE, PhD, MA, BSc, ARCS, FRIC.
A barrister, Founder and Vice President of MENSA, the society for people with exceptionally high I.Q. A Council Member, former Chairman and Vice President of the IPI. He also held office with IFIA.

Mr Andrew Warren, IEng, LicIQA, AIPDC.
The inventor of **'SLIDE-SAFE'** ™, the auto-safety device which makes sliding patio doors much safer, especially when children are around. 'Slide-Safe' is mounted on the door-frame above children's reach. A spring-loaded wheel automatically comes into play as a buffer once the door is open a small amount. The device comprises a fixing housing containing a short spring-loaded pivoted arm carrying a flexible rotatable cylindrical shock absorber and two finger pulls. When the door is slid open the pivoted arm swings the shock absorber against the closing jamb of the door frame. A safety gap is created which ensures that little

fingers don't get crushed or trapped. The device has won several invention, design and safety awards during the past few years, and it has recently been selected for **Millennium Product status** by The Design Council, which clearly recognises the importance of this product. ROSPA, (The Royal Society for Prevention of Accidents), welcomes and supports Slide-Safe as a contribution to child accident prevention. After vital support from The Federation of Small Businesses, he was able to get a DTI grant to progress the product to the market. He is looking for overseas licensees, and has international patent and design protection. Andrew lives in Wolverhampton, trading as Waverley Design & Engineering Services.

Mr Michael Watson.

A Council Member for some years. An **Inventrepreneur** with much background in design and development of electromechanical equipment, mainly from a self-employed base. He is particularly known for his invention of the '**WATSON-LINC**', ™, a telephone recording device which permits interaction with the bi-directional currents in a number of cored cables. This is achieved without actual electrical contact despite the signals being buried in common mode noise. This telephone recording method does not require plugs or sockets. It has a security aspect and has been successfully commercialised in partnership with Retell (a company specialising in telephone recording). Mike is confident his other inventions coming down the line will be successful too.

Mr Ken White.

Techmart Inventor of the Year in 1984, for his **'COMPAK (Strawbox)**' invention. Following the 1973 oil crisis, agricultural industry was undergoing a chronic shortage of packaging materials and prices rose. Ken conceived the idea of taking raw fibrous material that by a simple process could be used as a replacement for wood. Compak Systems Ltd was set up for the purpose. They now make a variety of single-use boxes and trays (non-returnable). Straw is resin bonded under heat and pressure to form board which is easily shaped. Simplicity of the process allows larger volume users to do it themselves in-house. There is worldwide interest.

Mr Eric Whittaker.

By the age of 12 he had built every model in the Meccano book. As a teenager was tea-boy to R. J. Mitchell of Spitfire fame at the Supermarine works in Southampton. At seventeen, Eric invented a cycle alarm and a hill-steepness indicator. Later, it was a cycle attachment for a school satchel, an engineer said he should patent it. He was one of a group of highly inventive youngsters, 2 of his pals died testing out home-made diving suits (one in deep water, the other in a bath). Next came a three-speed hub for a cycle, with only half as many parts as a Sturmey Archer. After the war he became a patents engineer with Normalair Ltd (related to Westland A .C.). Eric worked for various aircraft manufacturers, but all the while knocking out his own inventions. Most well-known for his preposterous suggestion in a letter to ***'THE INVENTOR'***, that ***inventors were most likely to be of the Gemini starsign and be about 29 years of age*** when producing their most significant work. This kept the IPI postbag full for months.

Mrs Iris Whyte.

Winner of the first 'Woman Inventor of the Year' award in 1998. Her '**TALKING POTTY**' invention is a toilet-training aid for the very young. A daughter had a young child with toilet problems, so Iris designed a potty with a microchip in the base, so that when a child had used the potty it talked to the child in a 'praise-reward' style that encouraged preferred habits.

Mr Alan Wilcher, IEng, MIAgrE, MIRTE.

A Member of the Board of Management of the Institute since 1988 (a Fellow). Also the UK Youth Advisor for IFIA. Founder of the **Association of Innovation Management (AIM)** in 1989, which was later subsumed to the IPI in 1998. He organised the **Radio Cambridge Inventor's Phone-in feature**. He also formed a branch of the IPI in Cambridgeshire. In the Rotary Club, he is active in 'Young Inventor/Designer' awards. A project engineer with much civil engineering and agromechanical experience, he has been a Consultant and troubleshooter for new product development projects. Alan successfully designed products for industry, including **'Top-flo' dryers** for Lely, **'Remarkabull'** oestrus indicator for Janssen, **'Yokel'** dairy feeder for RECO, 'Pic-fix' photo-retouch kit for MPP, **WSA 3-D animation** for Technic. He is still consulted regarding earlier projects. He now has his own company 'Inventurous Ltd', developing new products and licensing for private companies. He has a particularly wide group of valuable contacts which makes his advice especially effective when advising inventors.

Mr A. J. Wilkinson.

A Fellow from 1930, by virtue of his new process for rubber production. **'LINA-TEX'** is a rubber made by his process, patented in Malaya. Vulcanisation takes place at atmospheric pressure and temperature, which means that any reinforcing materials used are not be subject to the weakening forces of heat. The end-product is doubly-resilient, abrasion-resisting and non-perishable. This material has a wide range of applications because of its excellent chemical resistance. His company had headquarters in Chicago.

Mr. Jeffrey Woolf.

The inventor of '**MICROMAP'**, trading as 'I to Eye' Ltd. This system of credit-card sized maps and data-cards inside a viewer has received a lot of acclaim at exhibitions, it won awards in 4 different sections at INPEX 1995 (USA). Later it won the 'Sunday Times' Invention of the Year award (1996). This device put an end to carrying around a bunch of bulky maps. Ordnance Survey is enthusiastic about Micromaps, and sells these products in their shops. Another of his inventions that was quite clever was called **'VISIGARD'** (for the motorist). It was a vehicle security device whereby after forced entry into the vehicle, the windscreen suddenly becomes opaque, so the vehicle cannot be driven. It is simple for the owner to install without professional help.

Dr Brian Wybrow, CChem, MRSC, PhD (Lond.).

A Member of the Institute for many years. In addition to being an inventor himself, he operates a Patent Consultancy Service and has partners with expertise in intellectual property matters and illustration/preparation of technical literature. In his early career he was an R & D scientist, but frequent exposure to chemicals which affected him necessitated a career change into innovation management. His recent inventions include a prismatic cassette case for audio-cassettes. An elongated transparent prism-shaped plastic bar inserts in a fixed manner into a cassette case lid. The description/title of the cassette contents are then visible through the thin edge of the case even though the writing is on another inner surface. Another item is a double-ended clamp, which is all-plastic and usable from either or both ends. There are no metal parts. So it is valuable for all manner of wires, cables, tubing. A very versatile device designed by Brian using CAD, then the drawings were used by Geoff Forward, Development Engineer at Ticona (UK) Ltd to create the product. There has been a need for a double ended-clamp without a separate spring and having no metal parts, Brian was the first to solve this problem. He has also been involved in the development of clockwork-driven devices for over a decade. In 1987 he started to patent a clockwork generator, and had prototypes demonstrating the conversion into electrical energy, from the potential energy of a clockwork motor. In the process he discovered a clockwork torch had been invented as far back as 1940. There were clockwork lamps on toy trains in 1909, clockwork-driven dynamos in 1934, and another clockwork-driven generator in 1942. He obtained a hand-operated electric generator that was made by Meccano in 1975. Brian then decided to withdraw his patent application, and was taken aback when the Baylis clockwork radio came on the market. Wybrow Innovations is based in Ruislip, Middlesex.

Protecting Your Idea

IMPORTANT

This section briefly outlines the various means of protection that can be used to protect against plagiarism. It is intended only to give a general review of the law. There are numerous exceptions and complications to the principles outlined here. Always seek professional advice on any legal matter.

The Institute of Patentees and Inventors does not offer professional advice on legal matters. It accepts no liability for loss incurred through reliance on information given in this document which is based on current UK laws and practices.

INTRODUCTION

It is very important for the inventor to understand that a potential licensee will not license an idea, only the rights to that idea. This means that the inventor must have some form of intellectual property rights or legal ownership that he or she may sell or assign to the licensee. If a person came up with the idea that "it would be a good idea to make an electric teapot", then this idea would not be able to be protected nor licensed. If that person then sketched, made a model or produced a detailed design for that electric teapot, then this further work could be protected and assume intellectual property rights which could then be licensed or assigned.

Some protection, such as copyright and unregistered Design Right, generally comes into force automatically when a new idea is created; but stronger protection, involving patenting or design registration, has to be applied and paid for. In some cases, more than one form of protection may be appropriate, for example to protect both the workings of a new idea and its appearance.

Some inventors/innovators decide not to spend time and money on protecting an idea, preferring to take a risk and get it into the market-place as quickly as possible and thereby generate both cash and a name in that market. However, it is still important to make sure that someone else's rights are not being infringed by this action. Anyone can check whether an invention has already been patented by paying for a computer search of existing patents at the British Library or at regional patent offices with on-line facilities. A patent agent will also carry out this search.

It is possible to take out insurance, both against being taken to court for infringement and against the costs incurred in taking action against someone plagiarising one's own idea. Inclusion of a patent on a Register of Insured Patents, held by an insurance company, can in itself be a deterrent against plagiarism.

COPYRIGHT *2-DIMENSIONAL DESIGNS*

Copyright is relevant to all original two-dimensional works such as text, drawings, paintings, engravings, photographs, posters, graphic works or similar artistic works. It protects the expression of the idea but not the idea itself. Copyright only protects what is recorded on the drawing etc. and it does not protect the general principle of the idea. This means that it would not be an infringement for someone else to use the same idea as long as the form in which the idea was expressed was not copied. It protects only against copying of work and not against the independent creation of the same work by another person.

Copyright confers on the owner the right to control the exploitation and other treatment of the original work. The right covers copying or reproducing, adapting, publishing. performing and broadcasting of the work or a substantial part of it.

When the work is created, copyright arises automatically and so does not have to be registered or paid for. The copyright exists but it would be very difficult to prove in a court of law the date on which that copyright was created. It is very good practice for any inventor to keep an invention diary that details all the steps taken in the development of their ideas. Also the inventor should regularly sign and date original drawings, photographs of models and any other similar information and then have these witnessed and dated by someone a court would accept as independent and reliable. This witness should obviously be someone the inventor trusts, but should also can be relied on to maintain the secrecy of those documents. The originals should also be copied so that the inventor always has a "back-up" set of information available. The witnessed originals should then be deposited in a safe place and treated as valuable

Copyright can be licensed or assigned outright to third parties, inherited and dealt with like any other piece of property. It lasts for the lifetime of the 'author' plus 70 years. However, where a two dimensional design applies to an article which is exploited commercially (for example, a textile design or a pattern to be applied to a teapot) the design is protected from reproduction on such articles only for a period of 25 years from when the article is first marketed. There are certain other exceptional cases and, as in all areas of intellectual property rights, professional guidance should be obtained if the inventor/author is in any doubt.

Copyright is extended automatically to all Berne Copyright Convention countries, which covers most of the world. Copyright protection is also extended automatically to all Universal Copyright Convention countries provided that the standard UCC copyright notice is placed on the article or design. The notice should consist of three elements, namely, the symbol © which is the capital letter C contained within a circle, the name of the copyright owner and then the year of first publication of the work (e.g. © Fred Bloggs 1995). Very often the words "ALL RIGHTS RESERVED" are added to the notice; this is actually required in most South American countries. The owner of copyright is usually the author of the work, but if the author is an employee employed to create the copyright work in the course of his or her duties, the employer will normally own it. Generally a freelance artist retains the copyright even in a commissioned work, though there are exceptions to this rule.

DESIGN RIGHT *3-DIMENSIONAL DESIGNS*

Before the 1988 Act, copyright covered unregistered three-dimensional designs in the same way as two-dimensional designs. There are complicated transitional provisions dealing with such designs, and copyright still applies for one-off works such as sculptures.

For three-dimensional designs such as a new teapot design first created after the new Act came into force, a new Design Right replaces copyright for the protection of most features of shape and configuration. Certain designs whose shape is dictated solely by the function they have to perform or whose shape must match or fit other articles are excluded from protection. Two dimensional pattern or ornament on the surface of the design will continue to be protected by copyright.

Design Right allows the owner to control the exploitation or other treatment of his or her original design. The right covers making articles to the same or substantially the same design, and dealing in them. As with copyright it protects an author only against copying of the design and not against the independent creation of the same design by another person.

The period of protection for Design Right is 15 years from the embodiment of the design in a design document or prototype, or 10 years from first marketing of the design, whichever is the shorter period; however, after the first 5 years from the start of marketing, licences to reproduce the design must be available to anyone prepared to pay a reasonable royalty.

Design Right arises automatically as does Copyright. A similar procedure as described in the section on copyright above can be used to gain the maximum protection.

There are no marking requirements but preferably a mark similar to that used for copyright should be adopted on such unregistered design works to maximise any protection which may be available overseas under copyright. (e.g. Design Right 1995 Fred Bloggs). The Design Right can also be licensed or assigned in the same way as a copyright.

Unlike the situation with copyright, freelance designers lose their Design Right to the person commissioning a design from them. Also there is no major international agreement on the protection of unregistered designs overseas.

DESIGN REGISTRATION *2 & 3-DIMENSIONAL DESIGNS*

This form of protection is available in aesthetically appealing designs which are either two-dimensional or three-dimensional. It applies to features of shape, configuration, pattern or ornament which appeal to the eye.

Protection is excluded for purely functional products and there is no protection for features whose shape is dictated by the need to match or fit other articles. In order for a design to be registered whilst it is still secret, it must be 'new' at the date of application for registration. In general a design is considered to be 'new' only if it has not been published, offered for sale or sold in the UK. Also it must not be the same or very similar to another design already registered in the UK, but in practice this test is not very severe. It is possible to seek protection for only that part of a design which is new if it adds a new feature to an old design.

Because of the requirement that the design must be new, you must apply for registration whilst it is still secret and before selling or offering it for sale in the UK. However, a design may be displayed at an exhibition certified by the Department of Trade and Industry without invalidating a subsequent application for registration, provided that the application is made within 6 months after the opening of the exhibition. It should be noted that very few exhibitions are certified by the DTI and so in practice, this provision should not be relied on.

Registered designs give the owner the right to control the exploitation and other treatment of the design. It gives a monopoly and so, unlike Design Right, protects an author against the independent creation of the same design by another and not merely against copying of the design.

Protection for registered designs filed before the 1988 Act is for a maximum period of 15 years from the date of filing of the application. Protection for registered designs filed after the new Act came into force lasts a maximum of 25 years. Fees are payable on application and on renewal of the design every 5 years.

Applications for design registration in the UK should be filed at the Designs Registry within the Patent Office in London or Newport. The applicant must be the author of the design or the person who, as employer or as the commissioner of the design, acquired ownership of it, or someone to whom right in the design has been assigned.
An applicant can also apply to register a corresponding design in a large number of foreign countries. These foreign design applications are backdated to the UK application date (known as the priority date) if they are filed within six months from the date of filing the application in the UK. If application is made beyond this period, it is very unlikely that the foreign design applications will be successful.

PATENTS

It is important to realise that the principle behind the patent system is that in patenting an invention, a patentee gives the idea to the public domain and in return for a fee, is given an exclusive licence to exploit that patented idea. Therefore if after starting the patent process and after the patent has been published, the patentee chooses not to continue the process or is unable to pay the fees, the patentee has no further rights in the idea and it is in the public domain.

A patent protects a new invention i.e. a technical development. An invention is patentable if it is novel, if it involves an inventive step and is capable of industrial application.

Certain things may not be patented. These include a discovery, scientific theory or mathematical method, mental process, literary, artistic or aesthetic creation, a scheme or method for doing a mental act, doing business, presentation of information, or a method of diagnosis or treatment practised on the human or animal body. In this last area, it is interesting to note the debate now raging about the practice of patenting isolated DNA and genes. The resolution of this debate could, in the longer term, change the interpretation of what can and cannot be patented.

To be novel, an invention must never have been disclosed publicly in any way, anywhere, before the date on which the application for patent is filed. If the invention has been disclosed, it will not be possible to obtain patent protection for it.

An invention must also be regarded as involving an inventive step. This terminology has been likened to the phrase "revolutionary rather than evolutionary". It means that, at the time of application, your idea would not have been obvious to a person skilled in the field. Numerous applications fall down on this criterion and innumerable cases are argued whether or not an invention is obvious.

For an invention to be capable of industrial application, it must be possible to make or use it in some kind of industry. This means that the invention must be at least theoretically usable in some form of apparatus or device, product, material or substance, or be an industrial process or method of operation.

The owner of the invention is the inventor or the person who as employer of the inventor acquired ownership of it, or someone to whom the inventor has assigned the right in the invention.

Applications for patents in the UK should be filed at the Patent Office in London or Newport. It is possible for anyone to apply for a patent by supplying a description of the invention and completing the application form, but professional advice should always be sought where possible, particularly if filing a complicated patent application.

The extra expenditure involved in using a patent agent is almost essential, mainly because the protection obtained depends on the wording used to describe the invention. The initial application can be a description accompanied by drawings, and a formal description with formal claims to the invention, defining in legal terms the invention for which legal monopoly protection is sought, must be submitted within 12 months. Alternatively the formal application can be filed immediately. There is currently no fee for filing a provisional application.

When the initial application has been received, the Patent Office gives the application a date and number. No further action is required by the owner for up to one year but this should be an extremely active year. The owner must decide whether to exploit the invention or not to take it any further, in which case the invention will not be published. If the application is to go ahead then the next step is a search by the Patent Office Examiner, for which there is an additional fee. The results of this search are sent to the applicant.

Eighteen months after the application is first filed, the Patent Office publishes the invention. A further payment, for the examination stage, must be made within 6 months of the publication if the inventor wants to continue, having considered the Examiner's search report. The Examiner will then consider all the details of the application and decide whether to grant a patent or not. If the patent is granted, there will be no more to pay until annual renewal fees are required from the end of the fourth year after the date of filing. The UK patent lasts for a maximum of 20 years from the date of filing

An applicant can also apply to register a corresponding patent in a large number of foreign countries. These foreign patent applications are backdated to the UK application date (known as the priority date) if they are filed within 12 months of the first application in the UK. It is usually not possible to file in foreign countries if this time scale is missed.

Filing the patent in other countries is expensive; it could cost between £800 and £2,000 per country. However, the 12 months 'breathing space' after filing the initial UK application gives the inventor time to decide whether this expenditure is worthwhile.

The UK is a member of international schemes for handling patent applications and is able to handle applications for the European Patent Convention (EPC) and the Patent Co-operation Treaty (PCT).
Both these schemes allow an inventor to apply, in one country, for national patents in other countries. The EPC covers 17 European countries and is operated by the European Patent Office. The PCT covers nearly 50 countries worldwide and is available through the local patent office. If there is a requirement to gain patent coverage in other countries as well as the UK, it may be that the EPC or PCT route may be more efficient. A patent agent will be able to advise you on this matter.

TRADE MARKS

A trade mark is an identifying sign used by a person, company or other organisation to distinguish their goods or services from those of other traders. The owner of the mark can apply to have the trade mark entered on the Register of the Trade Marks Registry. This can be done either before starting to use that particular mark or afterwards. There are 42 classes in the Trade Marks Register and an applicant needs to be able to identify which class or classes are appropriate.

A trade mark is registerable if it is a distinctive means of identification of the goods or services claimed, for example, a newly invented word, a new symbol or design. Marks can also become distinctive if they are used substantially in relation to particular goods or services. Marks which consist of geographical names, or common surnames, or words which are purely descriptive of the goods or the character of the goods in question, are not usually registerable, at least not without evidence of distinctiveness through use of the trade mark. Marks can be assigned and licensed to third parties.

The owner of a registered mark can sue for trade mark infringement. Infringement results from unauthorised use of a trade mark and where confusion or a loss of reputation may result. Professional advice should be sought if infringement is thought to be occurring.

A trade mark is initially protected for 7 years and is then renewable every 10 years. The registration can be kept in force for an indefinite period provided that renewal fees are paid, but may be removed if it is not used for a continuous period of five years.

Applications for trade mark registration in the UK should be filed at the Trade Mark Registry at the Patent Office in London or Newport.

Under the terms of an international convention, an applicant can also apply to register a corresponding trade mark in a large number of foreign countries. These foreign trade mark applications are backdated to the UK application date if they are filed within six months of the first application in the UK. Unlike Patents and Registered Design, trade marks can still be registered in foreign countries after this six-month period.

PASSING OFF

The law about Passing Off protects against a person or company misleading or deceiving the consumer as to the origin of the goods, products or services by offering them in a very similar get up or design to one that is already on the market and has an established reputation. It also prevents the use of a trade name or mark on similar products or services to one which is already used by another trader who has acquired a reputation and goodwill in the name, thereby confusing the public as to the origin of the goods. This offers some protection to owners of unregistered trade marks with an established reputation.

TRADE SECRECY

Prior to the publication of an invention, design or other work it is possible to protect it by the laws of confidentiality. Put simply, if the work is submitted to a person by its owner and that person is made aware that the work is being shown in confidence, that person may not use or disclose the work without the consent of the owner for so long as it remains unpublished. Once the work has been published by or with the consent of the owner, or has come into the public domain without disclosure by the person to whom it was shown, that person cannot be restrained from disclosing it or using it unless a copyright or other right can be relied on by the owner.

A trade secret is often the "black art" that a company uses to manufacture a product. It might be the special moulding technique used or how an intricate form is created. The company does not publicly protect these techniques but maintains strict secrecy so that competitors do not learn these trade secrets.

ABBREVIATIONS USED THROUGHOUT THE BOOK

AIM	Association of Innovation Management (merged 1996)
BOT	Board of Trade (now the DTI)
CIPA	Chartered Institute of Patent Agents
DTI	Department of Trade and Industry
EPC	European Patent Convention
GBIIF	Great British Inventions & Innovations Fair
IFIA	International Association of Inventors Associations
INPEX	Inventions & New Products Exhibition (in Pittsburgh, USA)
IP	Institute of Patentees (1920)
IPDC	Intellectual Property Development Confederation (merged 1998)
IPI	Institute of Patentees & Inventors (from 1956)
LINPEX	London Inventions & New Products Exhibition
NIC	National Innovation Centre
NRDC	National Research and Development Corporation
PCT	Patent Convention Treaty
WIPO	World Intellectual Property Organisation
WW1	World War One
WW2	World War Two

The Institute of Patentees and Inventors

For further information regarding membership benefits,
contact the Institute at the address below.

To order additional (post-free) copies of "A *Bit* of an Idea",
send a cheque for £6.99, or telephone/fax/e-mail the Institute
with Visa or MasterCard number and expiry date
and your name and address.

Orders outside the UK, please remit £10 per book requested,
payment by International Money Order, Visa or MasterCard only.

The Institute of Patentees and Inventors,
Suite 505a, Triumph House,
189 Regent Street,
London
W1R 7WF

Telephone: 020 7434 1818
Facsimile: 020 7434 1727
e-mail: ipi@invent.org.uk
www.invent.org.uk